WEATHER AND CLIMATE

in colour

WEATHER AND CLIMATE
in colour

Svante Bodin

Illustrations by
Studio Frank

BLANDFORD PRESS
POOLE   DORSET

First published in England 1978
English language edition copyright © 1978 Blandford Press Ltd
Link House, West Street, Poole, Dorset, BH15 1LL

World copyright © 1978
Almqvist & Wiskell Förlag AB, Stockholm, Sweden.

ISBN 0 7137 0858 1

British Library Cataloguing in Publication Data
Bodin, Svante
   Weather and climate. – (Blandford colour series).
   1. Meteorology
   I. Title
   551.5   QC861.2 .B621 1978

ISBN 0–7137–0858–1

Text printed in Great Britain by
Richard Clay (The Chaucer Press) Ltd, Bungay, Suffolk

CONTENTS

1 INTRODUCTION

'Some are weatherwise, some are otherwise.'

Benjamin Franklin

The American journal *Weatherwise* has taken this quotation as its motto and there is still a lot of truth in what Benjamin Franklin said. The person who understands weather *is* wise.

In earlier times, weather was not merely a topic of casual conversation, but a matter of life and death. Indeed, in many places in the world today it determines the vital factor in the feeding of your family for another year with a successful crop, or facing the consequences of a drought or flood.

The atmosphere can release enormous power in the form of tropical revolving storms which frequently threaten the coastal regions around, for example, the Gulf of Mexico or the densely populated coastal regions of Asia. Strong winds destroy houses, torrential rains and storm surges cause flooding and put whole regions under water. Each year hundreds of people die in typhoons, as the tropical cyclone is called in Japan, or in hurricanes, as they are called around the Gulf of Mexico. In middle latitudes, polar depressions often disrupt the day-to-day life of society. In northern countries, such storms produce heavy snowfalls during the winter, paralysing communications, blocking highways and isolating parts of the country for several days. These Atlantic low-pressure areas are usually accompanied by strong winds, sometimes reaching up to hurricane force. Ships along the coasts of Europe have to remain in harbour. Crews of fishing boats usually recognize the signs of an approaching storm, but they also receive warnings from the weather services.

During the summer, intense thunderstorms often occur. The up- and down-draughts in a thundercloud create enormous differences in electric charge between different parts of the cloud, and between the cloud and the ground. These charge differences are then released as bolts of lightning. Very often, thunderstorms occur after a long period of dry weather and they can start extensive forest and prairie fires. In the American Midwest the much feared tornado occurs which is an intense whirlwind which can pull up trees, lift cars and cattle high up in the air and smash houses to dust. The damage so caused amounts to several hundred million dollars each year and many people lose their lives. The worst outbreak of tornadoes ever recorded occurred on April 3 and 4, 1974, when more than eighty tornadoes were observed. The little town of Xenia in Ohio was practically wiped off the map.

The weather can be very dramatic, but most of us will probably never see a tornado. For most people the weather is something which only seems to be important when planning our leisure activities, such as a fishing trip or an outing to the seaside at the weekend. Although the rain is unwelcome during our few weeks of vacation, in some parts of the world the opposite may be true; in India, for example, where millions of people are dependent on the summer monsoon rains for growth of their crops. If the summer monsoon is late, a catastrophe lies just around the corner. In many parts of the world, water is the most precious substance of all and the weather is the major supplier of water.

It is easy to see how knowledge of the weather, its consequences and perhaps most important of all the ability to forecast the weather may be of great value. Even when it is not a matter of life or death, forecasts and an understanding of weather patterns can save large sums of money. Governments all over the world have realized this and in most countries there are national weather services providing warnings, forecasts and climatological information. People who are engaged in weather-sensitive jobs or activities have learned to work with the weather. Pilots seek information about weather on their route before take off. Farmers listen intently to the long-range forecast during harvest time or during the spring to find out the best time to sow. In many places, knowing just when the first night's frost will occur can be hard to predict and this is where the meteorologist can help the farmer and grower. Warnings of approaching storms can help everybody. For instance, snow clearing vehicles can be alerted, and people

2

may be able to close their shutters and lock up the house prior to the onset of a hurricane.

There are also other ways in which we can use our knowledge of the weather. We know quite well that there are variations in the amount of rainfall from time to time in any one place, but if we measure the annual rainfall this total is fairly constant from year to year. The same is true for the average annual temperature for a certain place. These average conditions, which are the results of weather patterns influencing a specific location, are studied in *climatology*. Other questions a climatologist may try to answer are: How often does it rain more than 30 mm during a twenty-four-hour period? How often does it rain a given amount in one hour? This information makes it possible for us to calculate the correct dimensions for water pipes, sewage and drainage systems, etc.

The mean temperature and annual rainfall are also helpful in deciding how well suited a particular place is for farming and which crops could best be grown. The climatological information tells us what we can expect, and it can also help the holiday-maker in his or her choice of destination by giving data on sunshine probabilities and typical temperatures.

Weather affects practically all aspects of our lives, whether we like it or not. The air we breathe must be clean and not too polluted. In our homes we want a steady temperature and a comfortable air humidity. If the humidity level gets too low, throats become sore or we start coughing. If the humidity is too high the room feels close and muggy.

Weather takes place on many different scales. The gust we feel when turning the street corner is only a couple of metres wide while the Atlantic storms, the polar front cyclones, involve distances of some 3000 km. We cannot see the largest weather systems directly. These are the jet-stream waves, the so-called Rossby waves, which have dimensions of 5000–10,000 km and circle the whole globe.

In the same way that we have different space scales, we also have different time scales. A gust only lasts for a few seconds. A thunderstorm may be active for a couple of hours, after which it starts to die. The low-pressure system with its fronts and snowstorms usually passes after a day or two. On the other hand, the long waves in the free atmosphere move only slowly and can affect the weather in a particular spot for weeks. How much does the climate change over the centuries? Why did the last ice age happen 10,000 years ago? These are time intervals which are difficult to comprehend, but they are time spans which the meteorologist must think about. Climatic changes have recently become the focus of attention in connection with the last three years of drought in the Sahel area in Africa. The field is very much open to speculation. Some people say that another ice age is coming, while others believe that mean annual temperatures will increase.

In this book we shall try to explain weather on all of these scales – both in time and in space. We shall also study climate and see how far scientists have come in understanding the reasons for climatic changes. This also opens up the fearsome possibilities of tampering with the weather and meteorological warfare. We have already used the word *meteorology*. Meteorology is the science which deals with the atmosphere and the weather. The meteorologist tries to discover the physical laws which govern the processes in the atmosphere and he also tries to apply this knowledge to predict the weather. To do so some of the most sophisticated products of modern technology such as giant computers, satellites and weather radar are employed. All are needed in order to analyse the complicated weather patterns of our world.

The atmosphere can concentrate its energy into huge outbursts. A polar front storm contains more energy than thousands of hydrogen bombs. Where does the atmosphere derive all this energy from? What source sustains the winds over the earth year in and year out? We now know that the sun is the ultimate energy source as it radiates huge quantities of energy towards Earth. But the sun rays are unevenly distributed over the surface of the planet, from north to south. The polar region only receives a small fraction of the energy that the tropics receive. An important function of the atmosphere is to transport this excess heat from the tropics to the northern parts, thus preventing a massive freeze-up. The storms of the middle latitudes play an important role in this process.

Yet the major part of the heat is not transported as heat that we feel with our skin or can measure with a thermometer – discernible heat – but in the form of heat which was used to evaporate water to vapour over the tropical oceans. When water vapour condenses again to water, this heat is added to the air as discernible heat. At the same time, the vapour in the air supplies us with the necessary rain, creates rivers and lakes, is stored in the ground and finally returns to the sea in an endless cycle. Water is almost as essential to weather as the air itself. In fact, a lot of what we call weather is merely water in one form or the other. This means that we must understand the weather in its entirety, and in order to do that we must also be able to understand the exchange between the oceans and the atmosphere. We must go even further than that. We shall see how human activity has already had a marked influence on the atmosphere and that threats to the atmosphere are increasing every year. We are gradually becoming aware that we are part of a huge ecological system which naturally comprises the atmosphere and the seas. If we tamper with one part in this complicated system, it can lead to adverse and unexpected effects in other parts.

We can consider the whole system as a heat engine, comprising the atmosphere and the oceans, and driven by the sun. This heat engine can show amazing diversity; it can give us beautiful summer days, or lightning flashes in the sky; in northern countries during the winter it can enshroud everything in a white cover of snow. It can be cruel and hard, but can also gently caress us with its warm spring breezes or let us lie on the grass on warm summer days creating imaginative figures out of the clouds. The heat engine works unceasingly to transport heat from the equator towards the poles, making life possible over a major part of the globe.

2 THE ATMOSPHERE

The air which we breathe is a part of the huge envelope of gases which surrounds the earth and which we call the atmosphere. Air is transparent, tasteless and odourless (if we haven't polluted it); despite this, however, its existence is obvious. The oxygen it contains is the basis for human life and, since air is in constant motion, we can feel the wind against our bodies. The air is also responsible for our verbal contacts as it carries sound waves and makes it possible for us to speak to each other. Without air there would be no sounds.

A most important ingredient of what we call weather is water vapour. Water vapour is also colourless, tasteless and odourless, and under normal conditions we cannot see it. However, when vapour condenses it forms fog or clouds, which are tiny water droplets or ice crystals suspended in the air. The clouds can give us rain and snow, hail and thunderstorms.

The composition of the atmosphere

Oxygen (chemical symbol O, but normally occurring as the molecule O_2) and carbon dioxide (CO_2) participate in the complex cycle called *photosynthesis*. Plants receive heat from the sun to use as energy, and take in carbon dioxide and water vapour from the air to produce the carbohydrates in the crop and the oxygen which is released into the atmosphere. The carbohydrates in plants are some of the most important components of our food. In fact, the whole oxygen content of the atmosphere was once formed as a 'waste product' of photosynthesis in the vast forests covering the earth several billions of years ago. When these forests became covered over and decayed under pressure, large deposits of oil and coal were created. In these deposits a substantial part of the carbon dioxide of the primitive atmosphere is bound up, as well as vast amounts of energy from the sun. We are using this energy now at an ever-increasing rate.

There is a tremendous amount of matter in constant motion around the earth. The mass of the atmosphere is about 5,243,000 billion tonnes. We have already mentioned some of the constituents of the atmosphere. The following table gives the most important gases and their amounts as a percentage of the total mass.

Principal gases in the atmosphere

Name of gas	Chemical symbol	Percentage of mass
Nitrogen	N_2	75
Oxygen	O_2	23
Argon	A	1·28
Carbon dioxide	CO_2	0·05 (variable)
Water vapour	H_2O	0·01–3 (variable)

A surprising fact shown in the table is the large amount of argon; 1·28% corresponds to around 66,000 billion tonnes in the atmosphere. Argon, however, is an inert gas which does not react chemically with other gases or compounds; it does not form molecules like oxygen and nitrogen, and because of these properties it does not have any important effects on the atmosphere. Besides the above mentioned gases there are a number of compounds in the atmosphere which can have important local effects although their percentage contribution is small. Taken over the whole globe, they represent a fair amount. Among these constituents we have two other inert gases, neon and helium, which exist together in an amount of over 70 billion tonnes. Another important compound is sulphur dioxide (SO_2), which may be the greatest air pollutant. The amount of sulphur dioxide in the atmosphere is about 10 million tonnes. Another very important gas is ozone (O_3), which is a variety, or allotrope, of oxygen. The amount of ozone is about 4 billion tonnes and the bulk of the ozone exists at heights of 15–50 km in the atmosphere. Electric discharges such as lightning, can also produce ozone in small amounts. After a storm ozone may sometimes be recognized by its characteristic sharp 'fresh' smell.

Water vapour and water

Besides oxygen, water vapour is the most important constituent of the atmosphere, yet, as shown in the table, the amount varies quite a lot. In the colder areas, as for example northern Siberia, the content of water vapour can be as low as 0·01%, but over the tropical oceans it rises to 3% at sea level. Water vapour is important because it can appear in all three physical states: ice (solid), water (liquid) and vapour (gas). We know the usual visual appearance of ice and water, but even these forms can take on strange appearances at times. Frozen water can appear as snow or hail. Snow can exist in a multitude of different crystalline forms. Water almost invariably exists as water drops in the atmosphere but their size can vary from a diameter of 0·001 mm in fog droplets to about 1 mm in rain drops. If the drops become bigger than 2 mm, they break up into smaller droplets. Water droplets found in clouds are usually of an intermediate size. Clouds can also appear in many different forms, depending on their altitude, method of formation and illumination by the sun. At an altitude of 6–10 km, the temperature is very low (usually below −40°C). At such a low temperature water drops cannot exist and are frozen into ice crystals. These give the clouds a special diffuse or fibrous appearance (cirrus clouds).

The most important feature of water is its ability to store heat when it turns into vapour. Let us consider a simple example. To heat a quart of water in a pot from room temperature to its boiling point at 100°C we need 340,000 Joules of heat (1 Joule = 1 watt second, Ws), or as much energy as it takes to keep a 100-watt light bulb burning for one hour. On the other hand, to evaporate this water to vapour without changing its temperature, we need 2,500,000 Joules or *seven times as much energy*! The amount of vapour existing in the atmosphere must have evaporated from somewhere; usually this takes place out over the seas, especially in the tropics. Huge amounts of energy are stored in the water vapour in the air. We call this energy the *latent heat of the air*. This heat can later be given back to the air when the vapour condenses to form clouds, so heating the air. This heating effect has a substantial impact on the formation of many weather phenomena, such as thunderstorms and tropical cyclones (hurricanes) as well as the common mid-latitude storms. In the same way that heat is required to evaporate water into vapour, it is also needed to melt solid ice into liquid water. That heat is about one tenth of the heat needed for evaporation; both are known in physics as 'latent' heats.

There is still another thing we should know about the water in the atmosphere. The amount of water which can exist as vapour depends upon the temperature of the air. Because water vapour is a gas, we can measure this amount by means of its pressure. Another way to state the maximum amount of vapour possible at a given temperature is by the mass of water vapour per cubic metre of air (usually grams per cubic metre or g/m^3). The maximum amount of water vapour the air can hold at a given temperature is called the *saturation value*. The actual amount held can, of course, be less than the saturation value, but rarely greater. (However, vapour can be transformed into liquid water by condensation. The saturation value only tells how much *vapour* is possible. Clouds contain both liquid water and water vapour.) The higher the temperature, the more vapour that the air can contain and the higher the saturation value. The following table gives a rough idea of how the saturation values vary. We see that grams per cubic metre

Temperature °C (°F)	Saturation value (g/m³)	
+40 (104)	40·0	(at normal
+30 (86)	30·4	atmospheric
+20 (68)	18·7	pressure,
+10 (50)	9·8	1000 mb)
0 (32)	4·9	
−10 (14)	2·9	
−20 (−4)	1·0	

(g/m^3) is an absolute measure of the content of vapour in the air. Another commonly used unit for measuring water content in the form of water vapour is the *Relative Humidity* (R.H.). The relative humidity tells us how much water vapour is present, expressed as the percentage of the maximum amount possible at the given temperature. At the temperature $+20°C$ we see from the table that we can have a maximum of $18 \cdot 7$ g/m³. If we actually had that amount at that temperature, we would have 100% R.H. If, on the other hand, we only had 10 g/m³, then our R.H. would be $\left(\dfrac{10}{18 \cdot 7}\right) \times 100 = 54\%$. We also see that 100% R.H. at $+20°C$ is $18 \cdot 7$ g/m³, but 100% at $0°C$ is only $4 \cdot 9$ g/m³! Now, if we have 1 m³ of air with a 100% R.H. at $20°C$, and cool this volume of air to $10°C$, then a certain amount of water vapour has to condense to liquid water. We see from our table that the maximum amount of vapour at $10°C$ is $9 \cdot 8$ g/m³. That means that the difference between this amount and what we had at $20°C$ ($8 \cdot 9$ g/m³) must condense, so that the amount of vapour at $10°C$ does not exceed its saturation value. The atmosphere has many ways of accomplishing this cooling, and the study of weather processes is largely a study of how temperature does actually change within it. Evaporation and condensation of water are two of the most basic and important weather processes.

Pressure, temperature and density

As we have seen, the atmosphere consists of a mixture of gases, of which oxygen and nitrogen are the most abundant. This mixture in essence remains unchanged up to an altitude of about 100 km. Below this height, the density of the air is great enough for the collisions that constantly occur between the gas molecules to keep the air well mixed. Above 100 km, the air gradually becomes less dense until its density is so low that the number of collisions that occur between the molecules can no longer keep the air well mixed. The composition changes, so that the heavier gases are found in a layer at the bottom and the lightest gases at the top. This means that, vertically, the atmosphere's composition gradually changes from oxygen, to nitrogen, and finally to hydrogen, which is the lightest of all elements. However, even this part of the atmosphere, where the air is extremely thin, is of particular interest because it absorbs the dangerous X-rays and gamma radiation from the sun. These rays would otherwise penetrate down through the atmosphere to the surface of the earth. However, this is a subject which is rather beyond the scope of this book.

The fact is that 90% of the *mass* of the atmosphere is contained in the lowest 15 km – and the layer in which all our weather is formed is about 6–8 km thick and comprises 66% of all the air. One can easily see that in relation to the dimensions of the earth, the atmosphere is a very thin envelope around our planet. The radius of the earth is 6370 km, while the atmosphere (from all practical respects) is about 15 km thick. Put another way, it is about $\frac{1}{400}$ of the radius of the earth. The atmosphere is like the skin around an apple. The horizontal dimensions of the atmosphere are much larger than the vertical ones.

Pressure

We can measure the air pressure at ground level with a *barometer*. Pressure is simply force per unit area per square centimetre; the unit in common use not so long ago was called an *atmosphere* (1 kg/cm^2). Nowadays meteorologists measure pressure in *millibars* (abbreviated as mb), 1 mb being $\frac{1}{1000}$ of a *bar*. One atmosphere corresponds to the pressure 1013 mb. Another common unit is *inches of mercury* or *millimetres of mercury*. This unit stems from the time when most barometers were mercury filled and one measured the height of a mercury column in an evacuated glass tube. Air pressure at the surface of the earth was normally capable of pressing the mercury in the tube up about 760 mm (30 in.) which corresponds to the mean sea-level pressure, 1013 mb. (In the metric SI units, or the *Système International*, the pressure unit is the *Pascal* which is the same as 1 Newton/m^2; 1 mb is 100 Pascal or 1 hektopascal.)

Temperature

Temperature is an important meteorological variable. Unfortunately, we also have different units for temperature. In some English-speaking countries, degrees Fahrenheit are still used while in most other countries degrees Celsius are used – often called degrees Centigrade. In this book we will use the term Celsius, and explain how to convert Fahrenheit degrees to Celsius, and vice versa. The construction of a temperature scale is based on finding two *fixed points*, where we define the temperature to be either 100 or 0 (zero). The Celsius scale, named after the Swedish physicist Anders Celsius, has the temperature at which water boils at normal sea surface pressure as +100 degrees. The freezing point of water is chosen to be 0. The Fahrenheit scale is based on the normal temperature of the human body as being 100 and the zero point is the temperature achieved by a mixture of ice and salt.

$$100°F = +37·8°C \text{ and } 0°F = -17·8°C.$$

To convert between Fahrenheit and Celsius

$$\text{Temperature in } °C = \frac{5}{9}(\text{Temp. } °F - 32).$$

There is also another temperature scale which is used mainly by scientists. That scale is called the *absolute temperature* scale and the units used are degrees *Kelvin*, after the English physicist Lord Kelvin. The zero point in this scale is the *absolute zero*, i.e. the temperature at which all motion ceases, all atoms and molecules come to a rest. This temperature on the Celsius scale is $-273°C$ or 0 degrees Kelvin which is written as 0 K (no degree sign being used). The freezing temperature of water, the zero point of the Celsius scale, is then 273 K. We simply get the Kelvin temperature by *adding* 273 to the Celsius degrees. For example, $+20°C$ is 293 K.

Density

All objects have a weight which depends on the gravitational pull of the earth. Everything also has a mass, which is independent of gravity and is an expression of the amount of matter in the object. In metric or SI units, this quantity is usually expressed in kilograms (kg). (Imperial units express mass in stones, pounds or ounces.) The more mass an object has, the heavier that object is. In order to compare different objects we need to know how much a standard volume of weight is found in different elements. In other words, how much mass there is in a cubic unit. In the SI system, we take one cubic metre and find out how much mass we have measured in kilograms. That will give us the *density* in kilograms per cubic metre (kg/m^3). On this scale of units water has the density 1000 since 1 cm³ of water weighs 1000 kg. Air on the other hand has a density of about 1 kg/m³.

The air pressure at the earth's surface is the weight of the air column directly above a given area. To show that this pressure acts in all directions, one can carry out a simple experiment. The only thing needed is a glass of water and a piece of paper large enough to cover the glass. Fill the glass completely with water, all the way up to the rim. Put the piece of paper on top of the glass and be sure that there is no air between the water and the paper. Now, carefully turn the glass upside down and slowly remove your hand from the paper. The water will stay in the glass as a result of the air pressure acting on the paper, thereby confining the water to the glass. However, make sure the experiment is done over the sink and be sure that the paper is absolutely tight up against the water surface, otherwise it won't work and the results will be somewhat dampening!

Hydrostatic equilibrium

If we go up through the atmosphere we will have less and less air above us, and so the air pressure will decrease with height. Correspondingly, in lakes and seas the water pressure increases with depth. Unlike water, air can be compressed and its compressibility is a complicating factor in the atmosphere. We know that pressure decreases with height, but what is the rate of this decrease? (This is an important point because most altimeters, the instruments in aeroplanes which measure height, are in fact pressure-measuring devices.) For example, how much does the pressure decrease over 100 m in height? We can get the answer by studying Fig. 1. We assume that the air is at rest. The slab of air is then affected by two forces. One is due to the gravitational pull of the earth. If our air volume is 10 m thick and 1 m² in cross-sectional area, and the density is 1·3 kg/m³ (which is normal at the earth's surface), the gravitational pull on the slab is 9·81 × 10 × 1·3, where 9·81 is the force of gravitation (g) per 1 kg of matter. But the air volume is also affected by pressure forces. In order for the slab to remain at rest, the lower part of it must be affected by a pressure force which is bigger than the one at the top

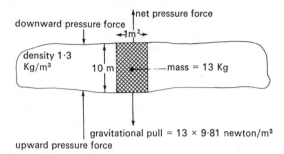

Fig. 1 A 'slab' of air.

of the slab. The net pressure force on the slab, the difference between the two, is then directed upwards and counteracts the
gravitational pull. In order for the slab not to move, these two
forces must balance each other *exactly*, i.e. the net pressure force
must be exactly equal to the gravitational force. If we carry out the
multiplications above, this is 127 Newton/m². Since 1 mb is 100
N/m², this means that the pressure difference, the pressure force,
must be 1·27 mb/10 m or 5 mb/40 m. The answer to our question
about the pressure decrease per 100 m is then about 13 mb. If you
live in a high-rise block you can easily check that with a simple
household (aneroid) barometer by measuring the pressure at the
ground floor and at the top floor and compute the difference.

We have already mentioned that air is compressible. This
means that the density is greatest at the earth's surface and decreases as one goes upwards in the atmosphere. But if density
decreases, then so also must the rate by which the pressure decreases. In other words, the rate of decrease is slower the higher
we go. Fig. 2 illustrates how pressure normally decreases with
height. We can see that the pressure decreases by 500 mb in the
first 5000 m, but only by 300 mb in the next 5000 m.

When the pressure force and the gravitational pull exactly
balance each other, we say that the atmosphere is in *hydrostatic
equilibrium*. In general, the air is in hydrostatic equilibrium. There
are some exceptions to this, for example, when we have strong
vertical motions in the atmosphere, such as occur in thunderclouds.

14

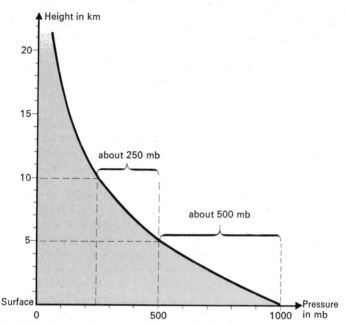

Fig. 2 The decrease of air pressure with height. The diagram shows pressure along the horizontal axis and height along the vertical axis.

Vertical stability

During the Napoleonic wars, at the beginning of the nineteenth century, there were rumours that Napoleon planned to invade England by means of hot-air balloons. He never got the chance to carry out this feat, but hot air balloons are still around and seem actually to have grown in popularity as a sport and pastime. The hot air balloon is based on the principle that if the pressure is the same for two gas volumes, the *density of the warmer gas is less than that of the colder gas*. So, if a balloon is filled with hot air, this air must be lighter than the surrounding air outside the balloon. The warmer air will make the balloon rise to the altitude where the two densities are the same. This is the same way in which a piece of cork rises to the surface when released under water. On the other hand, if we have a mass of air which is colder than its surroundings, its density is greater and so it is heavier than the surrounding air. This makes the cold air sink, something most people have observed when opening the refrigerator door; the cold air pours out on to your feet. The same principle operates in the frozen food cabinets in supermarkets, where the cold air is kept in place within the cabinet because of its greater weight.

Now, if we have a small mass of air moving upwards, for example in a hot air balloon, the pressure will decrease. When pressure is decreasing the air will expand. This expansion leads to decreasing temperature or cooling. This is something one can observe easily by placing a finger close to the valve of a car tyre at the same time as letting out some air. The air leaking out is considerably cooler than the ambient temperature of the surrounding air, despite the fact that the air within the tyre was at the same temperature as the air outside the tyre. The drop in temperature arises from the fact that when air escapes, the pressure decreases and the air expands when the valve is released. A temperature change which does not depend on any heat being added or removed from the air, as in this case, is called an *adiabatic process*. In the same way, we can have an adiabatic heating by rapidly compressing the air. How great is this temperature change, especially for an air mass moving upwards in the atmosphere? For dry air there is a simple answer: *the air mass will cool 10°C per 1000 m*. This temperature change with height is usually called the *dry adiabatic lapse rate*.

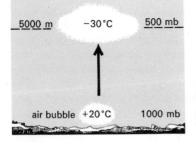

Fig. 3   In an air bubble, rising in the atmosphere, temperature decreases by 10°C per km. If the temperature at the ground is +20°C an air bubble lifted from the ground to a height of 5 km will have a temperature of −30°C

If air which, at the surface, has a temperature of 20°C, is forced to move up to an altitude of 5000 m, it will then be cooled to a temperature of—30°C (Fig. 3).

We can now consider three cases as illustrated in Fig. 4:

Unstable	*Stable*	*Neutral*
more than	less than	10°C/1000 m
10°C/1000 m	10°C/1000 m	

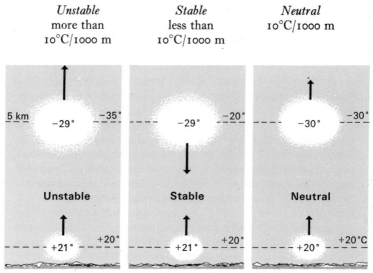

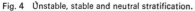

Surface

Fig. 4   Unstable, stable and neutral stratification.

In the first case, the temperature of the ambient air at the earth's surface is 20°C and only —35°C at 5000 m. This means that the temperature decreases more than 10°C/1000 m in the atmosphere. We now consider our small air mass again. Let us assume that it has been heated to a temperature slightly higher than the other air at the surface, e.g. by being over a warm piece of rock or over a hot black road surface which absorbs heat better than the environment. This air bubble will then be lighter than the surrounding air and will start to move upwards, in the same way as the hot air balloon. But while it is moving upwards it will cool at a rate of 10°C/1000 m. At the same time, the ambient air is becoming colder, at a rate of *more* than 10°C/1000 m. This means that the temperature in the rising air mass will always be higher than that of the surrounding air and, therefore, the density will be lower and the air lighter. The air will go on moving upwards at an accelerating pace. At 5000 m the temperature of the rising air will be −29°C as compared to −35°C for the ambient air, i.e. 6 degrees warmer. The vertical velocity, after having moved upward, can be 10–20 m/sec. Updrafts like this, or *thermals* as they also are called, occur on warm dry summer days. Gliders make use of them to gain altitude.

If the ambient air temperature decreases by more than 10°C/1000 m, any air which gains an initial upward motion will go on gaining altitude with an accelerating speed. We say that the air is unstable (unstable stratification).

Case two is a situation in which the temperature in the ambient air decreases at *less* than 10°C/1000 m. Let us assume that the temperature at the surface is the same as in the previous case, but that at 5000 m it is −20°C. In this case, air moving upwards will also be cooled at a rate of 10°C/1000 m – so at 5000 m its temperature is −29°C. However, in this case the temperature will be less than that of the ambient air and its density will be greater. The rising air is then heavier than the surrounding air and so it will sink downwards again. (Actually, it would never get as high as 5000 m, but would stop at a much lower altitude where the temperatures are roughly equal.) In this case, the temperature, decreasing at less than 10°C/1000 m, acts as a block against vertical motions. All air being forced upwards will be slowed down and will return to its original position.

When the ambient air temperature decreases at less than 10°C/1000 m, any air which receives an initial upward motion will be slowed down and come to rest. The air is stable or stably stratified.

The third case is the special case, when the lapse rate of the ambient air is exactly 10°C/1000 m. Air forced upwards will neither slow down nor be accelerated, but will continue to rise at the same speed as at the beginning. *The air is neutral or neutrally stratified.*

The vertical structure of the atmosphere

The three forms of temperature lapse rate we have studied here are all encountered in the real atmosphere. However, the stable stratification is the most common. Colour plate 'The Atmosphere 1–11' shows the vertical variation of temperature, illustrating the average conditions over an extended period of time over the whole globe; locally, and at specific times, it can appear rather differently. However, some of the features are constant in the atmosphere and they provide the basis for the way in which the atmosphere is divided into different horizontal layers.

The layer stretching from the earth's surface up to altitudes of about 11 km is called the *troposphere*. This layer contains about 90% of all the mass of the atmosphere and it is here that all the weather phenomena that affect us take place. The troposphere will be our main concern for the rest of the book. As one ascends through this troposphere, the temperature decreases at a rate of 6°C/1000 m on average; which is considerably less than the dry adiabatic lapse rate, showing that the atmosphere is normally stably stratified. The temperature continues to decrease up to an altitude of about 11 km, where it stops decreasing and remains constant at additional heights. When the temperature doesn't change with height, we have what are called *isothermal conditions*. (*Iso-* means 'the same' and it is used as a prefix in a number of terms in meteorology, as we will see later.) The level where we have the kink in the temperature curve, going from temperature decrease to isothermal conditions, is called the *tropopause*. The tropopause separates the (lower) troposphere from the (upper) *stratosphere*. The stratosphere is characterized by the fact of the temperature first remaining unchanged, and then starting to increase with height. This increase of temperature continues up to altitudes of 50 km, where the temperature reaches a maximum of +10°C. The temperature has thus increased by a total of 65°C through the stratosphere. The level where we find the maximum temperature is called the *stratopause*, which separates the stratosphere from the next layer, the *mesosphere*. In the mesosphere the temperature decreases again to its lowest level at a height of about 80 km. The temperature around the summer *mesopause* can reach values as low as $-120°C$ to $-130°C$. In the mesopause are the *noctilucent* clouds. These can be observed at northerly latitudes during the bright summer nights, when the sun never gets far below the horizon. Above the mesopause we find the *thermosphere*, which extends all the way out to interplanetary space. In the thermosphere, the temperature increases again, in direct proportion to the distance from the earth. X-rays and gamma rays from the sun are absorbed here and it is this absorption that causes the heating of the thermosphere (see colour plate 'The Atmosphere V–VI') to a temperature of some several thousand degrees. However, this cannot be compared with ordinary temperatures because the density of the air at these levels is about *one million times lower* than at the surface. Temperature here is really only a measure of the mean speed of the molecules. There is virtually no air in which, for example, to put out one's hand to feel the temperature.

The ionosphere

This method of classifying the different parts of the atmosphere is based upon the vertical temperature variation. If we look at the atmosphere from an electrical point of view, there exists another classification and this has been in use for a long time. The communications engineer or the radio amateur is interested in the *ionosphere*, which starts at about 70 km and extends for a couple of hundred km out into space. In the ionosphere, the air is more or less ionized, that is, there exist free electrons and positively charged particles. The ionization is generally concentrated into layers, designated by letters, of which the E and F layers are the most important, for the propagation of radio waves across the earth. These layers are affected by variations in the number of sun spots and by other activities like solar flares – violent magnetic 'storms' on the sun.

These disturbances on the sun produce showers of electrically charged particles which hit the earth's atmosphere after crossing space. The particles are of high energy and since they are charged, they are affected by the earth's magnetic field. The particles are 'steered' by the magnetic field lines towards the Poles. This concentrates the particles in particular regions, which are known as the *Van Allen belts*.

Aurora Borealis

The outbursts from the sun can cause visible phenomena, such as the aurora borealis (Northern Lights) and the *airglow*. South of 40°N, the aurora is rarely observed, but it is very common in northern latitudes. In places located around the earth's surface at about 75°N, one can observe the aurora practically every cloudless night (of course during the summer it is impossible to observe it by the naked eye because the midnight sun makes the sky too bright). The aurora borealis is probably caused by high-energy electrons colliding with air molecules, thereby exciting them in visible parts of the spectrum. The aurora is a fascinating spectacle of green, blue, red and yellow coloured curtains, moving up and down over the black polar night sky. The auroral display takes place at altitudes higher than 100 km, and much is still to be learned about how it is created. Similar displays are seen in Antarctica, being known as the Southern Lights or Aurora Australis.

Mother of Pearl clouds

Other beautiful phenomena are the *Mother of Pearl clouds* consisting of ice crystals which occur at heights of some 30–35 km. The name is well deserved because their faint iridescent light, in a meshlike structure, changes into different colours. They are most frequent over Scandinavia and Alaska in winter and are easiest to observe shortly after sunset or before sunrise, when the sun is illuminating them from below.

3 RADIATION

The earth receives almost all its energy from the sun, heat coming to us in the form of radiation. All the energy we use (except nuclear, geophysical and tidal energy) has its origin in the sun. Oil was formed from decaying vegetation over a hundred million years ago when the plants were using the sun's radiation in photosynthesis to produce the hydrocarbons of which oil is mainly composed. When we build dams or hydroelectric power plants we are only using water that was once evaporated by the heat from the sun and later returned to earth as rain in the mountain areas. The sun drives the huge weather machine both directly and indirectly. However, the energy we receive from the sun is distributed differently over the earth, causing various areas to have different thermal climates.

Electromagnetic radiation

What is radiation? We actually speak of different types of radiation, such as X-rays, radio waves, sun rays or light. Heat radiation or so-called infra-red and ultra-violet rays are other types of radiation. However, they all have one thing in common: they are examples of *electromagnetic radiation*. Electromagnetic radiation is present all the time around us. As its name suggests, electromagnetic radiation consists of electric and magnetic phenomena. Electromagnetic radiation is a wave motion where electric and magnetic fields successively increase and decrease in strength, oscillating back and forth like a string on a guitar. This oscillation is very fast; for example, ordinary radio waves can make about 100 million oscillations *per second*. The number of oscillations per second is called the frequency. At the same time this wave motion is propagated forward in space as the waves on a lake or the sea. The speed with which all electromagnetic waves move is 300,000 km/sec. With this speed it takes only eight minutes for the radiation from the sun to reach the earth which is 150 million km away. The type of radiation we have depends on the wave length or the frequency, which are related to each other. The wave length is the distance between successive wave crests. For radio waves the wave length varies between a couple of decimetres (dm) to several km. Visible light has wave lengths between 0·004 mm and 0·008 mm. In order to deal more easily with these short distances a special unit called an *Ångstrom* (*1 Å*) has been introduced, named after the Swedish physicist. 1 Å is $\frac{1}{10,000,000}$ of a metre (or in mathematical notation, 10^{-10} m, also called a *nanometer* in the SI system). The wave length for visible light using this unit is 4000–8000 Å. The shorter wave lengths between 100–4000 Å make up the ultra-violet radiation. Even shorter waves are called X-rays with wave lengths of 1–100 Å. The shortest waves are gamma radiation, with a wave length shorter than 1 Å. They are formed only in nuclear reactions, on earth or in the sun. Both gamma radiation and X-rays are dangerous to man in excessive doses. Carefully used, however, both have found important medical applications. Ultra-violet radiation is the one which gives us a suntan; excessive ultra-violet radiation can, however, give rise to skin cancer.

When light, or any electromagnetic radiation hits an object, one of several things can happen. Some objects are transparent like glass and simply let most of the radiation through. Our eyes are sensitive to light, most sensitive to green light, the wavelength which is absorbed the least in water. But a window pane does not let all the radiation through. Ultra-violet radiation is blocked out by glass so we cannot get suntanned behind a glass pane. Other substances are opaque and can take up the radiation and transform it into heat instead; they *absorb* the radiation. Black surfaces absorb especially well while white surfaces usually absorb very poorly. That is why people in subtropical or tropical countries often tend to dress in white clothes and paint the outside walls of their houses white to keep as cool as possible under the burning sun. Some substances can throw back most of the radiation which hits them without absorbing any or absorbing very little. Such objects are *reflectors*; a mirror is a simple and well-known example of a light reflector. Snow is also a very good reflector, a fact only too noticeable to skiers. Out in the snow one can quickly get a tan on one's face, but the reflected ultra-violet radiation can also lead to snow blindness if one is not careful to use sun glasses.

The sun's radiation

Before radiation from the sun reaches the surface of the earth it first has to pass through the atmosphere. The atmosphere, consisting of gases and a small amount of microscopic particles (aerosols), greatly influences the radiation from the sun as it travels down towards the earth. Colour plate 'The Atmosphere V–VI' illustrates what happens to the incoming radiation. The sun radiates energy in all wave lengths, from radio waves to gamma radiation, but the bulk of the energy lies in the visible part of the spectrum and in the infra-red portion (heat rays). The temperature of a radiating body can be determined by noting in which wave length it has its energy maximum. The higher the temperature, the lower is the wave length of maximum emission. For example, if we heat a piece of iron we can first see it turning red when the temperature starts increasing, and later changing to blue-white when the temperature exceeds about 8000 K. The radiation maximum of the sun corresponds to a surface temperature of about 6000 K. All bodies warmer than the absolute zero $(-273°C)$ radiate electromagnetic radiation, the amount increasing with rise in temperature. The energy maximum is displaced towards shorter and shorter wave lengths with increasing temperature. We will see later that the earth also radiates. Most of the sun's radiation disappears out into space, but at the earth's distance from the sun the intensity of the solar radition is still as great as 1400 watt/m² (the *solar constant*) which is engough to keep 14 of our 100 W lamps burning. The curves in Fig. 5 illustrate first how the energy is distributed in the different wave lengths when it impinges on the atmosphere and then what is left after passing through the earth's atmosphere down to the surface. At great heights in the air, higher than 100 km, practically all X-rays and gamma radiation are absorbed; this absorption causes the high temperatures in the thermosphere.

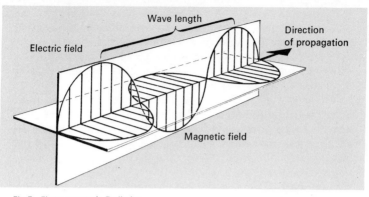

Fig.5 Electromagnetic Radiation.
Light, radio waves and X-rays are different examples of electromagnetic radiation, which consists of pulsating magnetic and electric fields. The wave length, which determines the type of radiation, is measured between two successive maxima in the electric or magnetic field strength seen in the direction of propagation of the waves.

The ozone layer

At 25–30 km above the earth's surface the density of the air becomes great enough for the ultra-violet radiation to start a series of *photo-chemical reactions*. Photo-chemical simply means that there are chemical reactions which need energy in the form of radiation to take place. Several photo-chemical reactions occur, but the net result of the impact of the ultra-violet radiation is that molecular oxygen (O_2) is transformed to ozone (O_3). Ozone, however, is not stable and decays to normal oxygen after a while. This decay reaction is promoted by the presence of oxides of nitrogen or chlorine. The amount of ozone depends on a balance between the rate at which ozone is created and the rate by which it is destroyed. The concentration of ozone also depends on the air motions in the vertical and horizontal directions in which ozone can be transported. Surprisingly enough the largest amounts of ozone are found at the poles during the spring months March–April. This is probably due to the general tendency of the air currents to transport ozone to northerly latitudes at heights between 15–30 km. The ozone needs ultra-violet radiation to form, but as soon as it has been formed it also absorbs a great deal of ultra-violet radiation; below about 15 km the amount of ozone in the air declines very rapidly. Ozone absorbs practically all radiation in wave lengths shorter than 3000 Å. Thus, the ozone layer acts as a very good shield against the intense ultra-violet radiation from the sun. However, some ultra-violet radiation does get through, but we have adapted to that by our ability to get a suntan.

The strong absorption of ultra-violet radiation in the ozone layer leads to the heating of that layer. Strongest absorption of radiation takes place at 40–50 km altitude and it is here we find the maximum in the temperature curve associated with the strato-pause.

Man and ozone

Small aberrations in the ozone amount can lead to more ultra-violet radiation penetrating through to the surface of the earth. Vegetation has also adapted to the normal level of ultra-violet radiation and an increase in ultra-violet radiation can lead to plants growing more slowly or even dying. The main danger to man is an increased risk of skin cancer.

Is there any possibility that the ozone might decrease and thus increase the ultra-violet radiation? Yes. Unfortunately, man has in his hands the means to damage the ozone layer. In 1974 two American scientists pointed out a group of chemicals called *Freons* as potentially very dangerous for the ozone layer! What is this chemical freon? Freon is the trade mark of a group of chemicals called chlorofluoromethanes (chemical formulae CF_2Cl_2 and $CFCl_3$). Freons are mostly used as propellants in spray cans and as cooling agents in refrigerators and freezers. They do not appear naturally in the atmosphere (or anywhere) and they are not re-moved by any natural process from the atmosphere once they are there. These properties mean that every time we use a spray can we add a small amount of freon to the atmosphere which *will stay there*.

The concentration of freon might then continue increasing up to values which could become harmful to the ozone layer. The production of freon is increasing 8% annually. Even if an extremely small part of the ozone layer is being affected right now, in a couple of decades the freon amounts might accumulate to a level where some scientists predict a 10% loss of ozone. This decrease would substantially affect the heating of the stratosphere, the air motion, the temperature at the surface of the earth and would result in higher doses of ultra-violet radiation. The active agent in freons is chlorine.

Another threat against the ozone layer is nuclear explosions, which produce large quantities of NO, nitrogen oxides. Since the explosions supply large localized concentrations they may be capable of tearing 'holes' in the ozone layer. There are study programmes under way to try to assess the effect of previous nuclear tests. There are indications that the effect on the ozone layer might be the most serious after-effect of a nuclear explosion.

Solar radiation and the troposphere

When the solar radiation has reached the troposphere the air density has increased to such an extent that the light rays begin to collide with the air molecules. This leads to *scattering* of the radiation. The amount of scattering is not the same for all wave lengths. It is greatest for blue light and least for red light. When the sun is high up in the sky the path for the rays through the atmosphere is comparatively short and only the blue light has 'time' to be scattered, so that the sky looks blue. When the sun is setting, the path for the light rays becomes longer and more of the other colours are scattered as well. Only the red light will penetrate down to the surface of the earth; the sun looks red and the morning and evening skies also look red close to the horizon. If there are clouds present in the vicinity of the sunset we can experience glorious colour displays. The scattered radiation reaches us as *diffuse* radiation, meaning that it comes from all directions. Since short-wave radiation is scattered more effectively than long-wave, ultra-violet radiation is scattered even more than blue light. Ultra-violet radiation is not visible, but we can see the effect of this process by the fact that we do not need to be exposed directly to the sunlight in order to get tanned. You can get tanned even on a cloudy day since a certain amount of ultra-violet radiation does penetrate the clouds, although the presence of clouds of course leads to reflection of some radiation back out to space by the cloud drops. The larger the fraction of the sky which is covered by clouds, the greater the reflected part of the incoming solar radiation. This reflection also acts like a scattering process, but in this case the scattering by cloud drops is the *same for all wave lengths*. This is always true when the scattering is accomplished by particles which are large when compared to the wave length of the scattered radiation. The light that reaches us on a cloudy day looks grey or whitish instead of the blue of a clear sky. The same is true for fog which gets 'milky' looking because all the wave lengths are scattered equally.

The character of the earth's surface varies from place to place; we have lakes, mountains, meadows and cities. Water covers about 60% of the northern hemisphere and 80% of the southern hemisphere. Land can be desert or be covered by forests. In the polar regions or during the winter in northern countries the ground is covered by snow. All of these affect the solar radiation when it finally reaches the ground. For example, snow reflects light extremely well. The ability of a surface to reflect incoming radiation is called its *albedo*. The albedo is expressed as the percentage of the incoming amount which is reflected. If the albedo is 0·55 that means that 55% of the incoming radiation is reflected back again. The albedo for snow is about 0·75. If we take an average over the whole earth we come up with a value around 0·32, taking into account the various coverage of forests, sea, snow, etc. We know that black surfaces absorb radiation very well; this also means that the albedo of black must be small. One can clearly see how a black asphalt road gets much hotter than the surrounding landscape during the summer. This property is also used practically in some parts of the world to speed up the melting of snow; a material such as lampblack is spread over the snow so that heat is absorbed more effectively thereby adding more heat to the melting process. Most of the incoming radiation which is not scattered or reflected back to outer space is used to heat the earth's surface. A minor fraction only is absorbed directly by the air to heat it.

Earth's long-wave radiation

We have mentioned before that all bodies radiate. Whether or not this radiation is visible depends on the temperature of the body. Our next problem is to find out the temperature of the earth and see how this affects the outgoing radiation. We now know that the average temperature of the surface of the earth as far as radiation is concerned is about $+15°C$. A body with this temperature radiates strongest at the wave length $13\ \mu$ ($11\ \mu = 0.001$ mm or 10,000 Å). At wave lengths longer than $50\ \mu$ and shorter than $5\ \mu$ the earth's radiation is negligible. The earth radiates in the wave length range called infra-red radiation, or heat radiation. The lower limit for infra-red radiation is at the red visible light. The amount or the intensity of infra-red radiation also varies with temperature and depends on the distribution of land and sea over the earth. Colour plate 'The Atmosphere VII–VIII' is a map which shows the outgoing infra-red radiation over the whole earth. The colours indicate the comparative radiation temperature. Since the temperature generally is highest in the tropical regions they also have the strongest outgoing radiation. The seas are generally colder than the surrounding land and radiate less, while the desert areas on the southern hemisphere have the strongest amount of outgoing radiation on the map. This picture is a compilation of radiation measurements from the American satellite NIMBUS 5 for the week of January 12–16, 1973. It therefore shows summer in the southern hemisphere and winter in the northern hemisphere. It is evident that the land temperatures are higher in the southern hemisphere than in the northern as shown by the radiation from, for example, Australia, South America and Africa.

Water vapour and carbon dioxide

The amount of heat leaving the earth's surface at the average temperature of $15°C$ is greater than the amount of incoming radiation from the sun. Since more heat leaves the earth than comes in, one would assume that the earth would consequently become colder. But we know that the temperature of the earth has been fairly constant or has only varied slightly since the last ice age. Something must prevent the earth's radiation from leaving the atmosphere. The missing links are *water vapour* and *carbon dioxide*. Both these substances absorb very strongly in exactly the wave length interval at which the earth radiates. All that is needed is a layer a couple of metres thick containing 0·03% carbon dioxide to stop completely all infra-red radiation at the wavelength 15 μ. For wavelengths longer than 18 μ and shorter than 8·5 μ water vapour is the most important absorber. Between 8·5 and 12 μ there is a 'window' which lets through most of the earth's radiation at this wavelength. Water vapour and carbon dioxide absorb radiation, but they also re-emit infra-red radiation. This is done as diffuse radiation in all directions. Half of this radiation is then radiated out to space while the other half is *radiated back to earth*. The resulting net loss of heat at the earth's surface is then considerably decreased by the presence of water vapour and carbon dioxide. It has been computed that if the earth lacked vapour and carbon dioxide, the average mean temperature of the earth would be −20°C instead of +15°C. Polar cold would then reign over the whole earth and life would hardly be possible. The atmosphere acts in the same way as a greenhouse; it is transparent to the incoming solar radiation, but prevents the heat radiation from escaping. This is called the 'greenhouse effect'.

The amount of water vapour varies markedly over the earth. In the northern regions the amount of water vapour is generally only about a tenth of what it is in the tropics. The surface then radiates more efficiently at northerly latitudes which results in a greater cooling effect. Snow is a very efficient source for infra-red radiation. Although white, it radiates almost like a 'black body', which is a term in physics given to a perfect radiator. The infra-red radiation during the long, dark polar night can make the temperature drop to very low values. The lowest temperature ever recorded at the surface is −88·3°C, recorded in Antarctica.

Clouds

In the same manner that the clouds reflect the incoming solar radiation they also reflect back some of the long-wave radiation from the earth. This is absorbed by the cloud water-drops and radiated back to the earth from the base of the cloud. During the day the influence of the clouds is of course greatest on the incoming solar radiation so that a cloudy day is often colder than a cloudless one when the sun can heat the earth without any hindrance. During the night when the sun has set, the lowest temperature during the night depends to a large extent on how much infra-red radiation is leaving the earth's surface. The cloud cover can stop this heat loss by radiation and the effect is that cloudy nights are much milder than cloudless ones. In northern countries during the long winter nights the difference between cloudy and cloudless nights can be as large as 15–25°C.

Radiation budget

What happens for the incoming and outgoing radiation is depicted in more detail in the colour plates 'The Atmosphere V–VIII'.

The intensity of the incoming solar radiation at the outer edge of the earth's atmosphere is 1400 W/m². But this is true only for an area positioned at right angles to the sun's radiation. The figure above shows that the incoming radiation per unit area is reduced the further north we go from the equator (or the further south). The total area of the earth is four times larger than the cross-sectional area that is facing the sun and we have to reduce the value above by a quarter to get the value the whole earth receives, i.e. 350 W/m². That is the value corresponding to the 100 units in the colour plates 'The Atmosphere V–VIII'. We see that of the 100 units at the outer limit of the atmosphere only 43% are left at the surface, namely 150 W/m². Colour plates 'The Atmosphere VII–VIII' show what happens to the outgoing radiation. The earth's surface radiates 115 units. Of this, however, the major part returns to the earth as re-radiation from water vapour or clouds. The net loss is 19 units. Until now we have ignored the heat which is transported away from the surface by means of evaporation, i.e. water taking up heat when evaporated to vapour, blown away by the winds and mixed by the rest of the atmosphere. Because of this constant mixing close to the ground (or turbulence – which we shall look at later in the book) discernible heat is removed from the surface; together another 24 units disappear from the surface in this way. This gives a net loss at the surface of −43 units which is exactly as much as the surface is receiving as solar radiation. Over the earth as a whole we then have a balance between the incoming solar radiation and the outgoing heat radiation and turbulent heat transport.

If we consider instead how the incoming and outgoing radiation vary from the pole to the equator we find that north of about 40°N (and correspondingly in the southern hemisphere of course) the earth radiates more than it receives while south of 40°N the incoming radiation is greater than the outgoing. The difference is greatest at the poles and at about 10°N. If the temperature conditions of the earth *only* depended on radiation, the equatorial regions would become warmer and warmer while the polar regions would become colder and colder. That, however, is not the case. The temperature differences between the poles and the equator start an air motion from the warmer tropics to the colder polar regions, in such a way that warm air is transported to the north and cold air is transported towards the south. In this way the temperature differences are smoothed out to a certain extent. Actually it is the temperature differences between the poles and

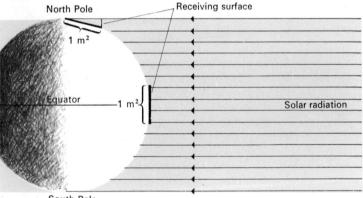

Fig. 6 Solar radiation reaching the surface of the earth heats the ground differently at different latitudes. At the equator light reaches the surface perpendicularly. At the poles an area of one square metre receives less light because the light rays hit the surface at an angle.

the equator which causes the atmospheric motion which leads to the storms of the middle latitudes and maintains the trade winds and the large calmer desert areas over the earth. The ocean currents also transport a substantial amount of heat from the south towards the north.

One is now getting an idea of how complicated the interactions are which take place between the motion of the air, water vapour, evaporation and condensation, rain, storms, solar radiation and the infra-red radiation from the earth. Water vapour and carbon dioxide are also important in understanding some of the problems in the field of climatic changes which have recently become of great concern to many countries.

4 WIND – THE AIR IN MOTION

The atmosphere is very rarely at rest. The motion of the air is what we call wind. We can have very localized effects like the flurry of wind at the street corner, or the sea breeze which is caused by the strong heating over land in late spring and early summer. Yet the wind can also be the air in motion in the large wind systems, anti-cyclones or polar front storms moving from the west to the east. Exceptionally strong winds occur in tropical cyclones, hurricanes and in tornadoes, where the wind speed exceeds 100 m/sec. Actually, no one knows for sure how strong the wind is in a tornado as most wind gauges break down in these violent conditions. Even regular mid-latitude storms can occasionally cause hurricane-speed winds posing serious threats to shipping and other human activities. Trees are blown down, bringing power and telephone lines with them, blizzards isolate villages, roofs are blown off buildings and scaffolding falls apart. Damage like this can run into vast amounts of money every year.

Winds can also be cold or warm. Northerly winds frequently bring colder air from northern latitudes, while southerly winds carry warm, tropical air. The winds and the organized wind systems are also important in understanding how the ever increasing, obtrusive air pollutants are transported in the atmosphere. Knowledge of winds has always been of great interest to seafarers and is now important with regard to modern air transportation; by avoiding strong head winds and making use of tail winds, a pilot can save both time and fuel. During the era of the sailing ships, the winds were not only part of the weather, but also the energy source for the vessels themselves. No wonder then that one of the most widely used wind-speed scales was developed by a sailor, the British Admiral Beaufort, during the nineteenth century. This scale does not in fact refer to speed, but to force – the effect of the wind on the sailing properties of a ship. (See pp. 252–253)

Wind speed and wind direction

To describe the wind we usually talk about the wind speed (though we really describe its strength or force) and the direction from which it is blowing. For example, a westerly wind means that the wind is blowing out of the west towards the east. The old Beaufort scale, even if it is still used in many countries over the world, has given way to other, metric, units like metre per second (m/sec.), knots (nautical miles per hour) or kilometres per hour (km/h). In the U.S.A., miles per hour is the commonly used unit. Quite often, we use names for different wind speeds, like 'breeze', 'fresh wind' or 'gale'.

Vertical wind

The winds described above are horizontal winds, blowing parallel to the earth's surface. But air can also move vertically and so we get a vertical wind, which can be directed either upwards or downwards. Up and down winds can sometimes become as strong as the horizontal winds. This is frequently the case in thunderstorm clouds. Normally the vertical wind is rather weak – much weaker than the horizontal. The vertical velocity in a large storm in the middle latitudes is of the order of 5–10 cm/sec. in the middle troposphere. This is in marked contrast to the horizontal winds in the same area, which can be of the order of 30–40 m/sec. Such a vertical wind is, however, strong enough to give rise to widespread areas of precipitation.

Why does the wind blow?

What keeps the atmosphere in restless motion? What causes the wind to blow? Since the wind is merely air in motion, there must be one or more forces acting upon the air, in the same way as an engine is needed to drive an automobile forward. In Chapter 2 we saw an example of a force capable of accelerating air vertically. For horizontal motion, the 'motive force' is the *pressure force*.

Pressure forces

Let us consider the following simple example (Fig. 7) where we have two points on the surface separated from each other by a difference in pressure. There exists a pressure difference between them of 10 mb. This pressure difference tends to move the air from the higher pressure to the point of the lower pressure. (By analogy, we need water pressure to get the water through a water pipe system to the taps or faucets. This is often achieved by building a tank inside the roof or a water tower with the water reservoir elevated. The pressure is then created by the earth's gravity acting on the water.) We can view the pressure difference in Fig. 7 in the same way. Where we have higher pressure the air tends to 'flow down the slope' to the lower pressure area. The pressure force between the two points increases with increasing pressure difference. Also, if the points are moved closer together, with the same pressure difference, the force also increases. The 'slope' increases. Normal pressure differences in the atmosphere are about 10 mb over distances of the order 500–1000 km.

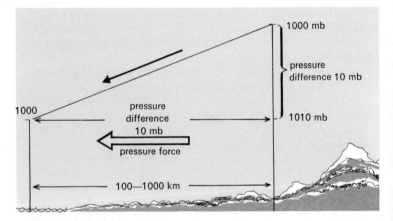

Fig. 7 A pressure force comes about when the air pressure is different at different locations. This means that there is more air where the pressure is high, and this excess air seeks its way to where there is less air, i.e. the pressure is lower.

The effect of the earth's rotation

If pressure was the only force affecting it, the air would move straight from the point of higher pressure to the point of lower pressure. If we draw lines for equal pressure on a map (these lines are called *isobars*, from *iso* meaning 'equal' and *bar* meaning 'weight') the wind would blow straight across the isobars (Fig. 8).

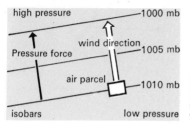

Fig. 8 If the pressure force was the only force acting on the air it would move directly from the high pressure to the low pressure, i.e. perpendicular to the isobars.

However, this is not actually what we observe, except, may be, close to and over the equator. What we see instead is the wind blowing parallel to, or almost parallel to, the isobars. The cause for this behaviour is to be found in the *rotation of the earth*, which changes the way in which we see the air moving in a very fundamental way. The impact of this is tremendous and the climate of the earth would have been very different if the earth had not rotated or had rotated with a different speed.

In order to clarify the picture and illustrate what is happening, we can try a simple experiment, using a turntable. Rotate the turntable so it moves counter-clockwise, the same way in which the earth rotates. Take a pencil or a piece of chalk and starting from the middle draw a straight line out to the edge. Because of the rotation a curved line is produced on the turntable which is deflected towards the right. In the same way, if we draw the line from the edge towards the middle, the line will again curve to the right. Although a straight line was drawn, for a person 'living' on the turntable and moving along with its motion (if one can imagine that), the line is curved to the right (see Fig. 9). The wind in the atmosphere moves relative to fixed lines in space, but we observe it standing on a rotating object, the earth. Suppose that

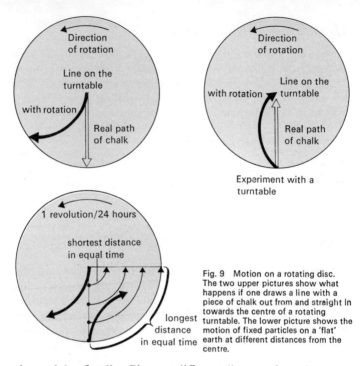

Fig. 9 Motion on a rotating disc. The two upper pictures show what happens if one draws a line with a piece of chalk out from and straight in towards the centre of a rotating turntable. The lower picture shows the motion of fixed particles on a 'flat' earth at different distances from the centre.

the earth is a flat disc. Places at different distances from the centre will move at different speeds when the earth is rotating. The disc rotates with one revolution every twenty-four hours, one revolution for all points. But during this time the places will have travelled different distances. Those closer to the edge of the disc will travel longer distances, resulting in higher speeds. To obtain a visual image of what happens in space, we can imagine a particle moving towards the middle, which will then have a greater speed in the direction of the rotation than the points it is passing over. This higher speed does not vanish simply because the particle is moving towards the middle, but will give it an apparant motion to the right. In a similar way a particle starting from the middle and moving towards the edge will have a lower speed than the points it is passing over and equally gain an apparent deflection towards the right. This effect is called the *Coriolis effect* or the *Coriolis force*, named after the Frenchman Coriolis who discovered it.

The earth is not a disc but a sphere. On a disc the speed with which a point is moving changes uniformly when going from the middle towards the edge; on earth this change occurs according to the latitude. The figure shows that at the equator the speed of a fixed point on earth is about 1670 km/h. At 30°N it is 1440 km/h, having decreased by 230 km/h from the equator to 30°N. At 60°N the velocity is down to 830 km/h and the difference between 60 and 30°N is then 610 km/h. Between 60°N and the pole at 90°N the velocity decreases to zero, a difference of 830 km/h. The effect of the rotation of the earth thus increases the nearer towards the pole we come. On the other hand, at the equator the change vanishes since two points a short way away from the equator on each side will have the same speed. *The deflecting 'force' towards the right on the northern hemisphere increases towards the north. It is greatest at the pole and zero at the equator.*

We are now in the position to imagine what happens when an air parcel starts to move towards lower pressure under the influence of a pressure force. As the air begins to move the Coriolis 'force' starts to act on it, acting at right angles to the motion and to the right, an effect which increases with wind speed. After a relatively short time the air will move in a direction *parallel* to the isobars, i.e. *perpendicular* to the pressure force, because by then

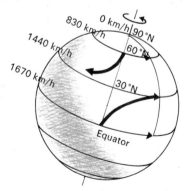

Fig. 10   The influence of the earth's rotation on the motion of the air.

we have a balance between the pressure force and the Coriolis 'force' which have become equal in strength and opposite in direction. The pressure force acts to the left of the wind direction and the Coriolis 'force' to the right. Such a wind, where the pressure force is balanced by the Coriolis 'force', is called *geostrophic wind* (geo meaning 'earth' and strophe meaning 'turn, twist') in meteorology. The Coriolis 'force' is proportional to the wind speed itself, meaning that the stronger the wind the stronger the Coriolis 'force', and the stronger the pressure force needed to balance the Coriolis 'force'. This can also be expressed as: *The stronger the pressure force, the closer the isobars, and the stronger the wind.* In fact, it is simple to derive a formula which will relate the wind speed to pressure force and Coriolis 'force'. For example, if we have a pressure difference of 10 mb over 1000 km at 45°N the wind is approximately 10 m/sec. or 20 knots.

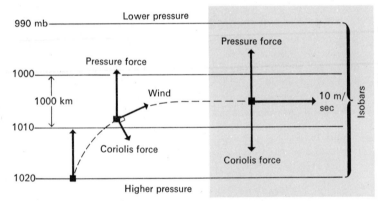

Fig. 11   A volume of air at rest will be affected by a pressure force so that it will tend to move towards lower pressure. When it picks up speed, however, it will be deflected towards the right because of the effect of the earth's rotation — the so called Coriolis effect. Finally a balance is reached between the pressure force and the Coriolis force and the air moves parallel along the isobars as is shown to the right in the figure.

(For the mathematically inclined we give the equation for the geostrophic wind as follows:

$$V_g = \text{geostrophic wind speed} = \frac{1}{fg} \cdot \frac{\Delta p}{\Delta s}$$

where

    g is the density of air. At the earth's surface about 1·3 kg/m³

    Δp is the pressure difference between two points

    Δs is the distance between these points measured perpendicular to the isobars on a map. $\left(\dfrac{\Delta p}{\Delta s}\right.$ is thus the pressure gradient$\left.\right)$

    f is the Coriolis factor which expresses the earth's angular velocity at a certain latitude. At 45°N f = 0·0001 s⁻¹.

With the aid of this formula the geostrophic wind can be calculated from weather maps. Generally isobars are drawn for the fixed interval of 4 mb. That means that Δp is always 4 mb or 400 Pascal. By measuring the distance between the isobars we can then compute the *pressure gradient* $\dfrac{\Delta p}{\Delta s}$ and thus V_g.)

We have also mentioned above that the Coriolis effect varies with latitude, becoming greater to the north. This means that a greater pressure difference is needed to balance the same wind speed.

The *real* wind in the atmosphere is often close to the geostrophic wind especially in the free atmosphere where the air is not influenced by the roughness of the earth's surface. The wind speed is, however, also influenced by the air moving in circular tracks, introducing a further 'centrifugal force', which is proportional to the square of the wind speed. Wind resulting from a balance between a pressure force, the Coriolis 'force' and a centrifugal force is called cyclostrophic.

Friction and turbulence

Close to the ground the air is affected by yet another force. At the earth's actual surface the wind velocity becomes zero (simply because the surface is not moving). In fact there is a gradual decrease of the wind when we approach the surface from above, for instance in an aircraft. The air is exposed to a *frictional force* which slows down the motion and 'backs' its direction, i.e. moves it anticlockwise. How high up does this frictional force affect the air? Well, that depends very much on the 'roughness' of the earth's surface. Houses, trees, rocks, hills and boulders force the air to move either over or around them. The effect is to create irregular eddies of moving air which cause mixing of the air in the lowest part of the atmosphere. This irregular mixing is called *turbulence*.

Most people have probably noticed that the wind rarely blows steadily, but that it changes its speed jerkily and irregularly. This is an expression of the turbulence of the air motion; particularly rapid changes are referred to as *gusts*. The wind speed we normally talk about and the one which is usually discussed by the weather forecaster is the mean wind speed, the mean usually being taken over ten minutes. The intensity of the turbulence or the gustiness depends a great deal on the roughness of the surface. It is greater over cities and mountains than over level plains or over the sea. The turbulence also depends on whether it is night or day.

As we saw in Chapter 2 the temperature in the air can decrease with height in three different ways giving stable, unstable and neutral stratification. During the day when the surface is heated

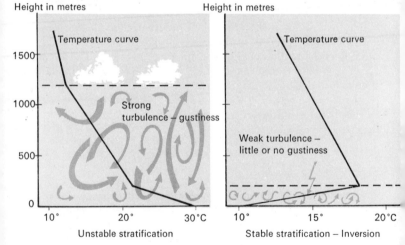

Fig. 12 Turbulence at stable and unstable stratification. During the day, when the earth's surface gets strongly heated by solar radiation, the vertical mixing of the air becomes very strong; large whirling eddies carry air up and down and the wind gets gusty. During the night or winter, inversions often form close to the ground. The vertical stability is large and the irregular turbulent motions are suppressed. The gustiness decreases or vanishes completely.

by the sun unstable stratification occurs near the surface and the roughness of the surface has a greater impact when forcing the air upwards. The air is more efficiently mixed and the turbulence increases. During the night the surface is cooled by long-wave radiation making lower layers of air more dense, and as the temperature increases with height (known as 'inversion') gives rise to stable conditions. Air parcels which accelerate upwards due to the roughness of the surface are slowed down by the stable stratification, so that the turbulence is suppressed or ceases to exist. The intensity of the turbulence also depends on the wind speed itself. The stronger the wind passing over an uneven surface the stronger the turbulence will be.

Strong wind in unstable atmospheric conditions can give rise to very strong gusts. If we have a mean wind of 15 m/sec. the gustiness under suitable conditions can amount to an additional 15–20 m/sec., i.e. the total wind speed in a heavy gust can amount to 30–35 m/sec. It is the heavy gusts which usually cause the most damage, and this kind of wind is especially tricky for sailors, since the gusts often come without much warning.

The turbulent layer close to the ground varies considerably in thickness. During the day when the turbulence is strong this layer can be 1–2 km thick, shrinking to about 50 m or less during the night when we have strong long-wave radiational cooling at the surface. It is especially important to study turbulence in order to understand how air pollution is diffused. Turbulence leads to mixing of the air, causing smoke from a chimney to be diffused both horizontally and vertically in addition to being transported downwind in the general air movement.

Turbulence and aviation

Turbulence does not only occur at the surface, but also occasionally in the free atmosphere. At altitudes between 5000–12,000 m a kind of turbulence called *Clear Air Turbulence* (*CAT*) sometimes occurs. This is a phenomenon well known to pilots and you may have encountered it yourself on a flight in a modern jet liner. Most of the time it only causes slight vibrations in the fuselage of the aircraft, but in exceptional cases the turbulence can get so violent that damage is caused to both aircraft and passengers.

Some crashes have also been attributed to clear air turbulence. When running into CAT, pilots usually try to alter their cruising level as soon as possible to avoid the CAT. To warn and forecast for CAT is one of the responsibilities of the aviation meteorologist. Very severe turbulence can also occur in connection with thunderclouds, so-called 'air pockets'. It is the very strong vertical velocities in the thunderclouds which act as turbulence. The up- and down-droughts can amount to 10–40 m/sec.; an aircraft can be lifted several hundreds of metres in only a couple of seconds and thrown down again just as quickly. These effects are worst at high flying speeds. Pilots try to avoid thunderclouds as far as possible and most aircraft are equipped with weather radar which can identify thunderclouds which are usually regions of severe turbulence. CAT is trickier since it occurs in clear air, as its name suggests and no satisfactory in-flight warning system exists for it.

Generally the turbulence decreases very rapidly above 1 km where the air is not affected by the frictional slow-down at the surface. Close to the ground, where we are, the friction, however, has a marked effect on the wind speed and the wind direction. The result is that the wind is weaker than it otherwise would have been from a pure balance between the Coriolis 'force' and the pressure force. The frictional force always acts in the opposite direction to the wind causing a slow-down in the wind speed. When the wind speed decreases the pressure force becomes stronger than the Coriolis 'force' causing the wind to turn towards lower pressure, instead of being parallel to the isobars (see Fig. 13).

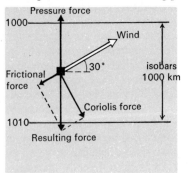

Fig. 13 Friction always acts as a retarding force on the wind. Air must stream towards lower pressure to counteract the frictional slow-down of the motion. A new equilibrium is established in which the friction force and the Coriolis force are balanced by the pressure force. The wind gets weaker as compared to frictionless flow and the angle between the wind direction and the isobars varies around 30°.

A balance is formed between three forces – pressure force, the Coriolis 'force' and the frictional force.

To sum up, the effect of the surface friction is that the wind speed gets smaller, the drop varying between 10–40% of the speed in the free atmosphere. The wind blows at an angle of some 10–45° over the isobars towards the low pressure area. The higher of these values is usually associated with stable stratification and the smaller value with unstable air.

'Backs to the wind' rule

Figure 13 shows that if we stand with the wind to our backs we have lower pressure to the left and higher pressure to the right. This is a well known saying which enables one to find out where the 'low' is, and is known as Buys Ballot's law.

The increase of the wind with height. Jet streams

In the lowest kilometre of the atmosphere the wind increases with height because the influence of the surface friction gets smaller until completely vanishing at about 1 km. But observations have shown that the wind often goes on increasing above that altitude up to 10–12 km where the wind speed becomes very great, in the order of 30–80 m/sec., but in extreme cases up to 150 m/sec. These strong winds, which are jets of flowing air, are concentrated in bands encircling the earth and are known as *jet streams*. The existence of these high wind speeds at 10–12 km in the atmosphere was discovered some years ago from observations of cirrus clouds which exist at these heights. But the final confirmation came during World War II when American bombers attacked Japan. Several pilots were met by suspicion and scorn when they told how they had missed their targets by several hundreds of miles, but more and more cases were reported and finally the only possible explanation was that of extremely high wind speeds. Nowadays we know that immediately over Japan the jet stream becomes unusually strong at these high levels. The map in colour plate 'Winds III–IV' shows the major jet streams in the northern hemisphere.

The increase of the wind with height is connected with the variation of temperature in the *horizontal* layers of the atmosphere! We have seen above that as the equator is heated and the poles are cooled a temperature difference between the two is created. But the temperature does not usually change gradually between the pole and the equator; the contrast is concentrated into special zones where the temperature can change by 10–15°C over a distance of 100–200 km. Such zones are called *fronts* or *frontal zones*. They will be discussed more extensively later, but in this context we can still see what effect such temperature contrasts might have upon the wind.

Figure 14 shows what happens when we have a pressure difference between two points at the same time as we have a temperature difference between them. At point 1 the surface pressure is 1000 mb while at point 2 it is 1010, 1000 km from point 1. As we have

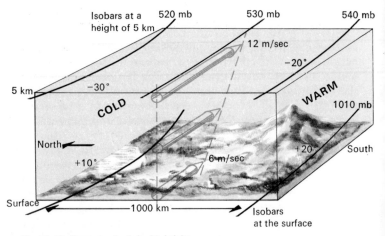

Fig. 14 The increase of wind with height.
Cold air has greater density than warm air. This means that pressure decreases more rapidly with height in cold air than in warm air. In the diagram there is a temperature difference of 10°C over a distance of 1000 km. Over the same distance there is also a pressure difference of 10 mb, being lowest where the temperature is lowest. If we go up to a height of 5 km the pressure difference has increased to 20 mb because pressure decreases more rapidly with height where the temperature is lowest. This also means a doubling of the pressure force. Because of this the wind has increased from 6 m/sec at the ground to 12 m/sec at 5 km. The wind increases because temperature varies in the horizontal!

seen this gives rise to a geostrophic wind of 10 m/sec. Now we also impose a temperature difference between the points of 10°C which is unchanged up to 5000 m despite the temperature decreasing with height. Let us assume that the temperature decreases by 8°/km giving $-20°$ and $-30°$ respectively at both points 1 and 2 at a height of 5000 m. Cold air is heavier than warm air, having a greater density. In the cold, heavier air the pressure will decrease more rapidly with height than in the warm, lighter air. This means that *the pressure difference between the two points will increase with height*. At 5000 m the pressure in point 1 is 522 mb while in point 2 it is 540 mb. The pressure difference of 10 mb at the surface has grown to 22 mb at 5000 m. Larger pressure difference leads to stronger wind as we have seen. *The wind speed increases with height* and the wind at 5000 m is about 22 m/sec. It is important to also realize that if the temperature difference had been reversed, with the colder air to the right, the wind would have *decreased with height instead*!

In the real atmosphere in the northern hemisphere the colder air is generally to the north of the warmer air. Since we very often have westerly winds at the surface, the west wind increases with height. The jet stream map in colour plate 'Winds III–IV' shows that the jet streams at heights around 10 km circle the whole globe and that the air moves systematically from the west towards the east. But we also see that there are displacements towards the north and south, so-called *jet stream waves*. In some places, particularly over the Atlantic, there are two jet streams, one at about 30–35°N and another positioned up at 55°N, which sweeps in over northern Europe. The southerly jet is called the *subtropical jet*, while the northerly one, which is associated with the temperature contrast between cold arctic or polar air and tropical air, is called the *polar front jet stream*. In some places they merge, forming one powerful jet stream as, for example, over Japan and south-east Asia and south-eastern U.S.A. The formation of the subtropical jet stream is a little different from the polar front jet. We will return to this subject later in the book. The jet stream map in colour plate 'Winds III–IV' shows the average wind speed at a height of 12 km. Since the position of the jet stream is not the same day after day the maximum wind speeds on an average map are slower than the wind speeds for a particular time.

Colour plate 'The Atmosphere III–IV' shows a cross section of temperature and wind in the atmosphere between the south pole and the north pole. As in the jet stream illustration, mean conditions are shown; the northern hemisphere shows those of winter and the southern hemisphere those of summer. The areas coloured red show the temperature distribution is warm and those blue that it is cold, with intermediary shades between. The highest temperatures are found around the equator while the lowest are found at the poles. The lines which separate differently coloured regions are *lines for equal temperature and are called isotherms*. They are numbered for every 5°C. Starting from the equator at the ground and moving upwards in the atmosphere one sees that the temperature continues to fall and we pass the isotherms for +15°C, −5°C all the way down to −85°C which is the temperature of the tropopause in the tropics. The tropopause is much higher over the tropics than over northern latitudes. Usually it is found at 16–18 km.

Now going from the equator to the left in the figure the temperature decreases, but we notice that there are kinks in the isotherms where the equatorial air changes into tropical air, tropical air into polar air and polar air into arctic air. These are the *frontal zones* in the atmosphere. One sees that the subtropical front does not reach the ground and that the arctic front only reaches up to about 2–3 km. Only the polar front extends through the whole troposphere. The average position of the polar front around the globe is at about 40°N during the winter (left side of the illustration) and around 50°N during the summer. However, large variations take place in the position of the polar front from place to place and time to time in connection with the polar front cyclones. If one moves further up in the atmosphere one notices that, when reaching 15–20 km, the temperature starts increasing with height again over the whole globe, reaching a maximum at 50 km, the stratopause. In the southern hemisphere where it is summer the temperature at the stratopause is about 10°C while in the northern hemisphere it is −15°C. Higher up we pass the mesopause and here one finds the lowest temperature in the hemisphere where it is summer, reaching the extremely low value of $-125°C$!

The illustration also shows the winds represented by *lines for equal wind velocity – isotachs*. The letters W and E indicate Westerly or Easterly winds (easterly winds are also indicated by putting a minus sign in front of the wind speed). These are average winds. The most important feature in the illustration is the wind maximum at about 12 km in connection with the polar and subtropical front. This runs true to the rule described above where the largest temperature change in the horizontal layer of the atmosphere must also have the largest increase in wind in the vertical layer. In this case the temperature decreases towards the poles, being strongest in connection with the fronts and giving rise to a strong increase in the west wind with height. There is a jet stream at 12 km from the west towards the east. This is equally true for both the northern and the southern hemisphere. Slightly north of 60°N at 25 km altitude we find another wind maximum called the *polar night jet*. At the pole during the winter night-time is continuous and because of the long-wave radiation the air continually loses heat, getting colder and colder. This gives very low temperatures in relation to more southerly latitudes and also an increase in wind with height. At the same height, but right to the south of the equator we find another jet maximum, only this time with easterly winds. This easterly jet increases up to an altitude of 60 km. In the northern hemisphere we here find an equally strong westerly jet. This illustration may seem rather confusing because it contains so much information, but it is a very good summary of many of the important points to bear in mind when studying the atmosphere, and is worth going back to now and again.

Low and high pressure

The two most prominent weather carrying wind and pressure systems in the middle latitudes are the low and high pressure systems. These may appear in many forms: as extensive areas of high pressure or minor migrating highs, intensive lows (depressions) in connection with polar front storms or extended low pressure areas, which can give unstable weather for several days over areas as large as half the North American continent or the whole of Europe. These pressure systems are also associated with characteristic wind circulations.

A *low pressure* area means a region where the *pressure is lower than its surroundings*. Similarly, a *high pressure* area is a region where the *pressure is greater than its surroundings*. If we draw *isobars*, lines for equal pressure, the low will show up as one or more concentric rings with its pressure decreasing towards the centre, called the low pressure centre, though in fact the isobars are not strictly circular. In a similar way the high will be a series of rings with its pressure increasing towards the centre, the high pressure centre. The air must move in such a way that pressure and Coriolis 'forces' balance, or almost balance, each other. This leads to an anti-clockwise circulation around the low pressure and a clockwise circulation around the high pressure (see weather map in colour plates 'Observations and Weather Maps III–IV'). The surface friction in the lowest kilometre of the atmosphere will deflect the motion of the air towards the centre of the low so that air accumulates at the centre of the low. The only way for the air to move then is up so that the inflow of air gives rise to an upward motion. An inflow of this sort is called *convergence* (of air). One can consider the motion of the air around a low pressure area as composed of two parts, one which makes the air flow towards the centre and one which circulates the low. In reality the air moves in spiral orbits towards the centre (see Fig. 15). The inflowing part is usually much smaller than the rotating part, but is large enough to create an upward motion amounting to 2–10 cm/sec. In the high pressure area the surface friction causes air to move out from the high pressure centre. This air is then replaced by some other air from above. Thus air is sinking in the central parts of a high; the outflow from the high is called *divergence* (of air) and gives a sinking motion of about the same order as the rising motion in the low. As pressure is simply the weight of the air in a column above a particular point, if air is flowing towards the centre of a depression that means that the amount of air in the centre of the low will be increased and the pressure in the centre should rise. If the low is to continue existing air must be removed from the central parts of the low at some higher level. For a low

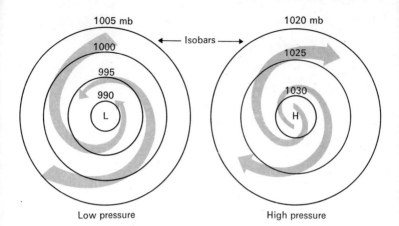

Fig. 15 The vertical air motion in lows and highs.
On a map with isobars lows and highs show up as a series of isobaric rings. The arrows show that air moves spirally in and out from a low pressure centre and a high pressure centre respectively, anticlockwise around a low and clockwise around a high.

which deepens or remains constant air is removed from its upper parts. At 6–10 km the low changes to a high with air flowing outwards. This outflow can persist without friction if the whole system, for example, is moving and this is often the case with mid-latitude lows. Correspondingly, the divergence close to the ground would cause the pressure to fall in the centre of a high if air was not flowing in at some higher level. In the upper troposphere or lower stratosphere the high changes to a low pressure area. The air motion then becomes as shown in Fig. 16. The rising motion in a low pressure area causes cooling of the air, and when the temperature gets low enough, condensation of water vapour to water drops, clouds and eventually rain, occurs. The weather in a low pressure area becomes unsettled and cloudy. In the high pressure area on the other hand the sinking motion makes the air warmer, the clouds dissipate and during the summer high pressure gives warm weather with sunny, clear skies. During the winter this might not always be the case, however. The weak winds in the high pressure area and radiational cooling during the longer nights can give rise to widespread areas of fog or

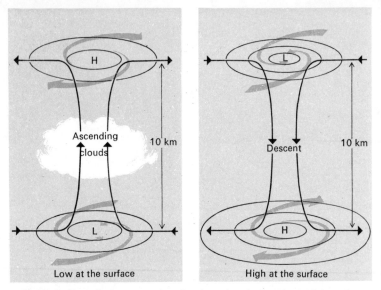

Fig. 16 In a low, air moves upwards and cools so that clouds and unsettled weather occur. In a high, on the other hand, the air sinks and heats, and dry and clear weather results. In the upper troposphere or lower stratosphere a low changes into a high while a high changes into a low.

low cloud; drizzle can also occur and light snowfalls or frozen fog are common in very northern latitudes.

We shall see, however, that even if this is true in general, the weather in connection with the migrating polar front cyclones is linked very closely to its structure in a very characteristic way. Actually the word *cyclone* is simply a word to describe how the air moves anti-clockwise as in a low. A high pressure is very often also called an *anti-cyclone* because the air moves in the opposite direction, i.e. clockwise.

Two other very useful terms which refer to the shape of the isobars and the flow of air are the (high pressure) *ridge* and the (low pressure) *trough*. The ridge is an extension of high pressure from a high. In contrast to a 'real' high, the pressure is higher than its surroundings in three directions only. The trough, as Fig. 17 shows, is an extension of low pressure from a low. The pressure

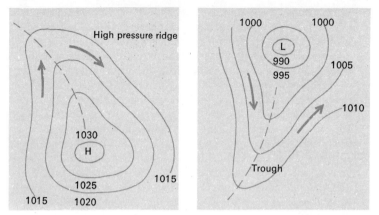

Fig. 17 Ridges and troughs.
The high pressure ridge is an extension of a high while the trough is a similar extension of a low. The weather associated with a trough is usually unsettled, with clouds and showers occurring. The ridge usually gives dry and clear weather as the high itself.

is lower than the surroundings in only 'three' directions as with the ridge. The weather in a ridge is very similar to that in a high and the weather in a trough is similar to that in a low. Ridges and troughs are seldom stationary, but generally move from west to east in the west wind belt of the middle latitudes.

Local winds – Föhn winds

In certain places around the world the wind can remain fairly constant in direction for a comparatively long time, such as a week or so. Such winds can dominate the weather and can carry warm air from the south so that one associates them with warmth and sunny skies, while other winds can be from the north and bring with them cold air from polar regions. Colour plate 'Winds III–IV' shows some examples of well-known winds in Europe. It is especially around the Mediterranean region that local winds are given traditional names, such as the Mistral or Sirocco. Of particular meteorological interest is the so-called Föhn wind, a dry and hot wind occurring frequently in the Alps, though the circumstances which give rise to the Föhn wind are found in many other places in the world. The inset in colour plate 'Winds III–IV' illustrates the formation of the Föhn wind. Humid and stably stratified air is forced to move up a mountain slope, cooling as it ascends. Gradually water vapour starts to condense to water, clouds form and it starts raining on the windward side of the mountains. When the air has passed the peak it starts descending again. When the pressure increases during the descent the air is heated. The clouds rapidly dissipate and the air becomes the same as it was before passing over the mountains, except for one thing: the latent heat released when the vapour condensed to water, clouds and rain, has now made the drier air warmer than it was before starting its trip over the mountains. This is a concrete example of the heat stored by water vapour and which can be returned to the air in the form of discernible heat. The temperature after the mountain passage can be 10–15°C higher than before, giving dry hot weather in the lee of the mountain range which lies across the path of the air stream. Some other places where this sort of wind frequently occurs are the Rocky Mountains (where the wind is called the Chinook, 'the snow eater', by the Indians), the Alps and the Scandinavian mountain range, and the Appalachians.

Land and sea breeze

Local winds can also be caused by differential heating over land and sea. Land and sea breeze circulations are well-known phenomena for people living along coasts or on large islands. Colour plate 'Winds I–II' illustrates this. We have already seen that warm air gets lighter than its surrounding air and rises, while cold air becomes heavier than the ambient air and sinks. If we have a vessel with a liquid, or air, and heat one end and at the same time cool the other end, the liquid will start to move. At the warm end the liquid will rise while at the cold end the liquid will sink. At the same time, however, the warm liquid at the surface will move from the warm region to the cold region to replace the liquid which is sinking. Close to the bottom the cold air will flow towards the warm region to replace the air missing because it is being heated. This is *closed circulation*. Since the only agents involved are cooling and heating, the circulation is called a *direct thermal circulation* (thermal has the same root as the word temperature, i.e. therme, meaning heat). Circulations like this are very common in the atmosphere.

The sea breeze is caused by the land at the coast being heated during the day by the sun. The sea has a vast capacity for heat and its temperature hardly changes at all by the solar heating during the day. The temperature difference between the land and the sea gives rise to a thermal circulation where cold air from the sea moves in over land where it is heated and rises. At a height of about 1000–2000 m we get a return flow bringing the warmer air back out over the sea where it sinks, so the cycle starts all over again. The effect on the pressure is also noticeable. When the air over land starts to get heated it gets lighter and the pressure decreases. The decreased pressure creates a pressure difference between the sea and the land and the air starts to move from the higher pressure area over the sea towards the lower pressure area over land. As we have already seen, the air is affected by the Coriolis 'force' after it has started to move. This makes the wind direction veer to the right during the day. On a coastline facing east we get an apparent impression of the sea breeze following the sun.

The sea breeze which during the afternoon can amount to 5–8 m/sec., dies out during the evening when the temperature over land starts to drop. During the night the land gradually becomes colder than the sea and the sea breeze during the day can then be replaced by a weaker land breeze blowing from the land towards the sea. In southern countries the sea breeze can often be pleasantly cooling. However, the sea breeze can also be treacherous to tourists from northern countries who want to get a suntan on holiday; the sea breeze makes people underestimate the strength of the sun at southern latitudes resulting in painful sunburn.

The stronger the temperature contrast between the sea and the land the stronger the sea breeze. This difference is usually greatest in late spring and early summer when the sea is still pretty cold while the land has started to get very hot during the day. The land breeze on the other hand becomes strongest during the late summer or early autumn when the water is warm and the longer nights cause greater radiational cooling over land. The sea and land breezes are also more prominent when there are weak winds and clear, sunny skies. That is often the case, as we have seen, in high pressure situations. The sea and land breeze can also be strengthened or weakened depending on the direction of the general wind.

Mountain and valley winds

Mountain and valley winds are formed in a way similar to the sea breeze. Colour plate 'Winds I–II' shows what happens during the night and the day in mountainous terrain. During the night the air along the mountain slopes is cooled, becoming heavier than the air away from the sides of the valley. The colder and heavier air flows down the slopes towards the bottom of the valley. We have a *mountain wind*. This effect is also evident on a smaller scale. When out walking or riding on a motor cycle in the late evening it is noticeable that the air in hollows and low-lying land is colder than air on the top of hills. This is also a result of the colder air flowing down into the bottoms of the hollows.

The opposite of mountain wind is *valley wind* which occurs during the day. As the illustration shows, the air along the slopes is heated more than its ambient air. The warmer air becomes lighter and moves up the slopes. In the middle of the valley the air sinks thus giving rise to a valley wind. In the same way as for the sea breeze, fairly undisturbed large-scale conditions are needed to produce a very noticeable mountain and valley wind circulation.

5 CLOUDS AND WEATHER

Before it was possible to measure the state of the free atmosphere using scientific instruments, clouds were the prime source of our knowledge of the atmosphere. Studying the clouds was an art which enabled people to make forecasts about the weather. This weather 'lore', based on centuries of experience, has been preserved in old sayings which we still use in conversation. They often contain a lot of truth.

Clouds can show the motion of air at higher altitudes, give information about strong vertical winds and whether the atmosphere is stable or unstable, i.e. if warm air is overlaying cold or vice versa. Many weather phenomena such as rain, snow or hail, thunder and tornadoes are formed inside clouds.

Clouds are, in fact, nothing more than water drops or ice particles, i.e. water in different forms. They can, however, exhibit a surprising variety of different shapes and they occur at all heights, from the surface of the earth up to a height of 80 km. Fog is simply a cloud that touches the ground. Between 6 and 10 km the cirrus clouds are found, which can reveal where the strongest jet streams are located. At 80 km above the ground are found the noctilucent clouds. Some clouds penetrate through the whole troposphere; these are very often associated with bad weather, rain or snow, in connection with fronts or cyclones.

Through the ages clouds have been classified by meteorologists and given latin names, based on those first devised by the British chemist Howard in 1803.

Clouds are divided into ten main cloud groups, each of which contains different varieties (see colour plate 'The Ten Main Cloud Types'). Before we knew how clouds were actually formed they were classified according to their shapes, for example *cirrus* means 'a hair' and *cumulus* 'a pile'; sheet clouds were called *stratus* (meaning 'a layer'). Clouds that occupy the whole depth of the troposphere are named *nimbus* (meaning 'giant'), such as *nimbo-*

stratus or *cumulonimbus*. *Stratocumulus vesperalis* form as left-overs from cumulonimbus clouds (*vesperalis* is derived from the latin word vesper meaning 'evening').

Although the latin descriptive names are used when reporting cloud formations, it is easier to discuss clouds in terms of how they actually form and the physical processes which are involved. The two main cloud types in this respect are *layer clouds* and *convection clouds*. Layer clouds form when a large volume of air is forced to move upwards and the air is stably stratified. This is generally the way that clouds are formed along the warm front in a depression. Air is also forced upward when it is flowing across a mountain range and *orographic clouds* form. On the other hand convection clouds form when the air is unstable and is moving upward in large bubbles in a jerky and irregular way. Such clouds get a cauliflower-like outline in contrast to the layer clouds which are smooth and without distinct contours and often cover the whole sky.

The convection clouds include the cumulus cloud and the cumulonimbus cloud. Weak convection and mild turbulence gives rise to stratocumulus and altocumulus clouds. At a height of 6 to 10 km cirrocumulus form when weak convection is present.

Cirrostratus and altostratus are examples of layer clouds, in the upper and middle layers of the stratosphere.

Water vapour turns into clouds

Dewpoint

Clouds form when the air is cooled down to its saturation point. At a specific temperature air can only hold a certain amount of water vapour. The *dewpoint* shows the temperature to which the air must be cooled in order to become saturated. If the air is cooled further, water vapour must condense to water. If one looks again at the table on page 9, one sees that if the air temperature is $+20°C$ and the water content is 9.8 g/m^3, the temperature must be lowered to $+10°C$ to achieve saturation. The dewpoint in this case is consequently $+10°C$. The dewpoint is always less than or equal to the ordinary temperature. If the air temperature is equal to the dewpoint the air is saturated and relative humidity is 100%. Dewpoint is a very convenient way of measuring the moisture content of the air since it shows directly which temperature is required for condensation and the formation of fog or clouds; the higher the dewpoint the greater the moisture content of the air.

Condensation

The fact that water vapour condenses at a relative humidity of 100% is only true if the vapour is in contact with a flat water surface, such as over a pond. In the atmosphere, however, the situation is more complicated. Water is present in the clouds in the form of small drops, their diameters varying between 0·001 mm to 0·1 mm. The water surface in such a small drop is very strongly curved. The large curvature decreases tension on the surface of the droplet which makes it easier for water molecules to escape from the water surface, i.e. evaporate. Water evaporates more easily from a curved surface than from a flat one. This shows that supersaturation, i.e. relative humidities above 100%, is required for droplets to form spontaneously by means of condensation. If there were no *condensation nuclei* in the air, clouds would be rarer than they are. In the air, however, there are large amounts of minute salt and dust particles which greatly assist the condensation of water vapour. The salt particles are of special importance since less than 100% relative humidity is required for condensation of water vapour over a salt solution. The salt and dust particles in the atmosphere are very small, their size ranging from $\frac{1}{10,000}$ mm to $\frac{1}{1000}$ mm. Ordinary 'clean' air contains 100 to 1000 particles per cubic centimetre while polluted air contains several millions of particles per cm³. The salt particles come from the seas where the winds catch the salt spray from the tops of breaking waves.

The condensation process starts on these condensation nuclei. As long as the drop is small the salt nucleus occupies a large part of the volume of the droplet and the salt concentration is high. The condensation can then take place at subsaturation, i.e. at relative humidities less than 100%. However, only when the relative humidity is above 98% do the droplets grow so large that they become visible as clouds. Smaller drops can be detectable as mist at a humidity as low as 70–80%, especially if the air is strongly polluted and contains many condensation nuclei. Highly polluted air is typical of large, industrialized urban areas. The air over large cities is very hazy and mist forms easily during the night.

As the drops grow larger the salt solution is diluted in the drops and the effect of surface tension becomes noticeable. At a size of about $\frac{1}{50}$ mm the drop requires a supersaturation of about 0·2% to get bigger. Cloud drops rarely exceed a size of 0·1 mm. Generally their sizes stop at about $\frac{1}{50}$ mm. Further temperature decrease and condensation instead produces more small drops. In fact raindrops with diameters of about 1 mm cannot form by means of condensation alone.

Cloud drops with a diameter about $\frac{1}{100}$ mm fall with a velocity of about 1 cm/sec. In the extensive areas with rising motion in a depression the vertical velocity amounts to 5–10 cm/sec. Cloud drops can therefore stay suspended in the air over long periods of time. If the rising motion dies out or decreases the cloud drops fall slowly downwards and evaporate so that the clouds clear. If the air sinks it gets warmed up and the clouds are dissolved faster. This is the case in the high pressure areas, which give clear and sunny weather.

Freezing

In the upper part of the atmosphere the air temperature is below freezing. The lowest temperatures are reached at a height of 10 km, down to −65°C. Pure water drops do not freeze at 0°C, but at considerably lower temperatures, about −40°C. Under normal conditions, however, there are impurities in the water droplet which act as *freezing nuclei*. They facilitate the transformation of water drops to ice crystals. In the atmosphere freezing normally takes place at temperatures between −10°C and −25°C depending on the amount and type of freezing nuclei. Water drops existing at temperatures below freezing are called *supercooled*. Supercooled drops are hazardous to aircraft as when they hit an aircraft, especially the propeller and the leading edge of the wings, they immediately freeze and form ice deposits which affect the lifting power of the wings. One of the tasks of an aviation meteorologist is to give warnings of icing conditions.

When air is saturated over a water surface it is supersaturated in relation to ice at temperatures below freezing. The largest super-saturation relative to ice is obtained at about −12°C. This is the reason that when freezing starts in a supercooled cloud the freezing takes place very rapidly. When ice is present the water vapour condenses directly to ice instead of forming water drops. The ice crystals grow rapidly and the water drops evaporate. In a couple of minutes a whole supercooled water cloud can change into an ice cloud.

Colour plate 'Cloud Physics' shows examples of drops of different sizes and ice crystals. In the upper right-hand corner the relative sizes of a condensation nucleus, a cloud drop and a rain drop are shown. Ice crystals can take on beautiful shapes and forms; they can be shaped like bars, plates, prisms and needles or look like stars. Snow flakes are composed of several ice crystals; they are very fragile and can easily break into smaller fragments of ice crystals or splinters, which in their turn act in the same way as freezing nuclei.

Formation of rain

As we mentioned earlier, cloud drops cannot grow larger than about $\frac{1}{50}$ mm by means of condensation. Drops of that size are too small to fall all the way down to the ground without evaporating. How is it possible then for a cloud to produce drops which may be a hundred or even a thousand times bigger than normal cloud drops? The Swedish meteorologist T. Bergeron put forward a theory in the 1930s which might explain how these big drops could be formed.

Cold clouds

The key to the solution of this problem lies in the fact that the tops of the clouds reach to such low temperatures that the cloud drops freeze. In contrast to water drops, ice crystals can grow very large at the expense of the water drops. Snow crystals then collide with each other and combine to form big snow flakes. When the snow flakes have grown large enough they become heavy and start falling. When falling they can collide with other snow flakes or supercooled drops thus growing even larger. When the snow flakes fall into the warmer, lower parts of the cloud they melt into big water drops a thousand times bigger than the surrounding cloud drops. On their way down they can collide with and pick up more water drops, as shown in colour plate 'Cloud Physics'. Rain drops vary in size between 1–2 mm; bigger drops usually split into smaller drops. The air stream around the drops slightly flattens them when they fall with a velocity of 1–2 m/sec. If the temperature is less than 0°C all the way down to the ground the snow flakes do not melt and the precipitation comes in the form of snow. Snow can occur even when the ground temperature is above freezing, provided the snow flake is falling rapidly enough not to melt before reaching the surface of the earth.

Warm clouds

In the tropics and particularly over the tropical oceans the air is so warm that all the clouds are at a temperature well above 0°C. In these clouds rain formation has to take place in another way. Water vapour condenses on condensation nuclei, e.g. salt particles. Over the seas there is an abundance of large salt nuclei. This fact makes it possible for some drops to grow bigger than other drops and obtain a higher fall velocity. When falling through the cloud the bigger drops collide with other smaller drops which adhere to them, so that they grow even bigger. This process is called *coalescence*. When they get big enough their velocities become greater than the upward motion in the cloud and they can fall as rain. Coalescence is also a contributing factor, as we have already seen, in iced clouds when big drops or snow flakes have already formed.

Convection clouds

Convection clouds are especially common during the summer when the ground is strongly heated by the sun; they form from rapidly ascending air bubbles. When these air bubbles move upwards they expand because of the decreasing air pressure and so become cooler. When the temperature has fallen sufficiently the air becomes saturated and water drops appear, thus forming a cloud. A necessary condition for this process to take place is for the air to be unstable (see page 18). The rather violent vertical motions associated with the upward motion of the air is called *convection* (simply meaning vertical motion). The most common convection cloud is the *cumulus cloud* (abbreviation Cu). To see how cumulus clouds are formed we can take a look at the upper panel in the colour plate 'Cb-Clouds I–II'.

Before the solar heating of the ground is sufficient to give the air bubbles a strong enough lift to move them upward to form clouds the vertical motions are referred to as *thermals*. Thermals are a well-known phenomenon to glider enthusiasts who use the uplifts in the air bubbles to gain altitude. The vertical velocities may amount to 10–15 m/sec.

How high does an air bubble reach before a cumulus cloud forms? That depends to a large extent on how humid the air is at ground level, but also how rapidly the air in the bubbles mixes with the surrounding air, i.e. how fast the air in the bubbles becomes diluted. In general, cumulus clouds appear at heights ranging between 500–2000 m, but if the air is dry they can form even higher up or even not form at all. Since the rising air must be balanced by an equivalent amount of air sinking in between the clouds only a fraction of the sky gets covered by clouds. Cumulus clouds rarely occupy more than half of the sky, usually less.

Cloud amount

In meteorology it is customary to measure clouds in oktas, i.e. eighths of total sky area. A clear sky is then 0/8 and overcast conditions correspond to 8/8. Half clouded is 4/8. Scattered clouds are counted by conceptually moving the clouds together and estimating the sky coverage.

The ordinary fair-weather cumulus has a fleecy appearance; these clouds are called *cumulus humilis*. Under favourable conditions, however, the fair-weather cumulus may increase in size.

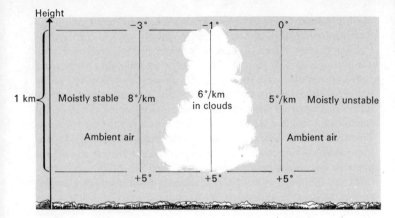

Fig. 18 When saturated air moves upwards in a cloud its temperature decreases by about 6°C/km, instead of 10°C/km as for dry air, because of the release of latent heat. If the surrounding air temperature decreases by *more* than 6°C/km the cloud remains warmer than the ambient air and can extend further upwards. The air is unstable when saturated or moistly unstable. If the air temperature decreases with less than 6°C/km the air is stable even when saturated and the cloud growth ceases; the air is moistly stable.

When condensation occurs in the cloud the heat of evaporation – the latent heat of the air – is given back to the air which is correspondingly heated. Because of this cloud temperature decreases at a slower rate than the dry adiabatic lapse rate, 10°C per 1000 m, the lapse rate being only about 6°C per 1000 m. If the ambient air temperature decreases with height faster than this the cloud bubble will continue getting warmer than the surrounding air and in that way remain lighter. The cloud can then go on growing upward. New protuberances stick out from the cloud in which the air is moving upward. (The second figure in colour plate 'Cb-Clouds I–II'.) The growth can be quite rapid. A big towering cumulus cloud is called *cumulus congestus* and looks like a giant cauliflower. When a cumulus cloud is able to grow in this way the air is conditionally unstable.

A cumulus congestus cloud occupies a large volume of air, the tops reaching some 3000–5000 m with strong upcurrents inside the cloud and vertical velocities of about 20 m/sec. On a summer day it can be interesting observing the growth of clouds, as in a matter of half an hour a small cumulus might grow into a mighty cumulus congestus.

The cumulus clouds contain more water than other clouds. The drops are relatively large but can stay suspended in the air because of the strong updrafts. In general the air surrounding the cloud is very dry and drops mixing with the ambient air rapidly evaporate. This gives the cumulus cloud very sharp contours. The liquid water content in a cumulus is about 1 g/m^3.

Cumulus clouds are most common during the summer. During the winter in northerly latitudes the sun is not strong enough to heat the air close to the ground to produce unstable conditions, but cumulus might form when cold air is streaming southward over, for example, warm seas and lakes, like over the Great Lakes or over the Atlantic Ocean.

Stratocumulus

When the wind is strong, the turbulence mixes the air vertically and moister air from the surface of the earth is transported upward. In such a case stratocumulus clouds might form at a height between 500 and 2000 m. Stratocumulus may also form when cumulus grow up into an upper inversion, a layer in which the temperature increases with height. Such a layer suppresses very efficiently the convective vertical motions and the cumulus clouds spread out horizontally into stratocumulus (Sc). Stratocumulus clouds can also form when the ground is heated only weakly by the sun, by, for example, the presence of higher clouds absorbing and reflecting the solar radiation.

Stratocumulus mainly consists of water drops and are common when fairly moist, warm air is flowing northward during the winter season. Stratocumulus are shown in the plate 'The Ten Main Cloud Types'.

Altocumulus

Altocumulus is a middle high cloud with its base varying between 2000 m and 4000 m. Altocumulus (Ac) can take on a variety of different shapes and usually form as a result of weak convection.

Sometimes they occur in cellular shapes, sometimes lining up in parallel bands spanning the heavens. Sometimes they may resemble stratocumulus. Mostly they consist of water drops but can occasionally contain ice crystals.

A particular kind of altocumulus is called *altocumulus castellanus*. Castellanus means 'tower' and they usually look like turreted castles and sometimes resemble cumulus. Altocumulus castellanus indicate considerable instability at the level where they occur, showing that fairly strong convection is taking place at higher levels. If observed in the morning during the summer, they are a warning that showers or thunderstorms may form later in the day.

Another rather distinguished form of altocumulus is the *altocumulus lenticularis* cloud that can be observed frequently in mountain regions. When stable air is forced up over a mountain the air flow takes on a wavy character behind the ridge. In the crests of the lee waves the air is getting cooled and as it becomes saturated clouds form. In the troughs the air has been heated as a result of the sinking motion and no clouds are present. They can be observed in the Scottish mountains, for example, and along the Scandinavian mountain range. This is shown in Fig. 19.

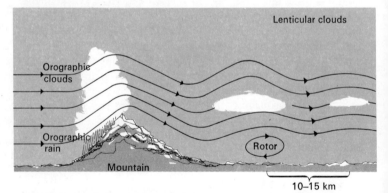

Fig. 19   Lee waves and Lenticular clouds.
When stably stratified air is forced over a mountain range, lee waves often form downstream of the range. Lenticular clouds form in the wave crests where the air is cooled to saturation. On the windward side of the mountain range the air is forced upwards, cools and gives rise to orographic clouds and precipitation.

Cirrocumulus

Cirrocumulus (Cc), occurring at heights between 6000–10,000 m, consists of ice crystals and is a high cloud. They are often called mackerel clouds and occur fairly rarely. Cirrocumulus show that the air at high altitudes can be unstable.

The clouds as we observe them do not stay unchanged and fixed. On the contrary they are constantly changing, forming, dissolving or transforming into other clouds. One cloud seldom exists alone, but in conjunction with other clouds at different heights. In the morning stratocumulus may break up into cumulus and in the evening when the sun-driven convection dies out cumulus are transformed into stratocumulus again. Towering cumulus congestus dissolve slowly when the sun sets leaving remnants of stratocumulus in lower layers and altocumulus in the middle parts of the troposphere. These evening clouds contribute to the beautiful play of colours at sunset and forecast another day with good weather.

Upglide clouds

The layer clouds are frequently associated with warm fronts, which carry with them most of the rain and bad weather in a depression or low in the middle latitudes. Layer clouds form when a large and extensive area of stable air moves upward. In contrast to convection clouds this occurs at a slow rate. The vertical upward motion takes place at a speed of only 5–10 cm/sec. as compared to 10–15 m/sec. in convection clouds. The slow rising motion produces clouds that have a smooth, diffuse and off-white appearance without details and sharp contours. Very often they cover the whole sky.

Orographic clouds

One way for the air to move upwards is when it is forced up over a mountain range (see Fig. 19). When the stably stratified air is forced up the slopes on the windward side of the mountain the air is cooled by expansion leading to condensation and cloud formation. Such clouds are called *orographic clouds* from 'oros' meaning 'mountain'. If the air is laden with moisture, rain and, higher up, snow also occur. This is the reason why mountain regions usually get more rain than normal flat land. Cherrapunji, on the southern slope of the Himalayas in India, averages 11,633 mm of rain per year!

A heavy, cold air mass can act as a 'mountain' if advancing warm and lighter air is forced up over the cold air. This is the case with warm fronts. Without forestalling the discussion about the polar front cyclones it seems appropriate to depict the development of layer clouds in connection with the passage of a warm front. In the colour plate 'Polar front systems III–IV' the typical warm front clouds are shown on the right.

Cirrus

Cirrus clouds look like long fibrous threads usually ending in a hooklike shape. Cirrus of this kind used to be considered the first sign of an approaching storm and are called *cirrus uncinus* ('uncinus' meaning 'claw'). They occur at heights varying between 6 and 10 km and consist of ice crystals (see colour plate 'The Ten Main Cloud Types').

Cirrostratus

A couple of hours after the first cirrus 'hooks' have been observed the clouds start getting denser and thicker until the whole sky is covered with contourless and transparent *cirrostratus*. It is sometimes possible to distinguish a banded structure in the clouds, but at other times the clouds are so thin that it is difficult to differentiate between them and polluted air. Cirrostratus, however, consists of a special type of ice crystal which causes the sun rays to refract and produce a bright ring around the sun, a halo. The halo phenomena is a definite indication that it is a cirrostratus cover we are observing. The halo has a radius of 22° and can be difficult to observe if the clouds are thin and the sun bright. By using sunglasses and covering the sun with one's fist it is usually possible, however, to observe the halo quite well. Cirrostratus occurs at the same heights as cirrus.

Altostratus

As a warm front approaches, the clouds thicken more and more. The halo disappears and the cloud base gets lower. The cirrostratus has been replaced by *altostratus* consisting of supercooled water drops. If the clouds are thin the sun can still be visible through the clouds, and appears as it would as if viewed through frosted glass. Thick altostratus clouds appear grey and smooth (see colour plate 'The Ten Main Cloud Types'), and are found at heights between 2–5 km.

Precipitating clouds

About six to twelve hours after the first cirrus clouds have been observed rain may start falling from altostratus, if it has grown thick enough. By now the cloud mass is extending through a major portion of the troposphere and the whole cloud is called *Nimbostratus* or the rain and snow cloud.

The upper parts of nimbostratus consist of ice crystals which are necessary for the formation of rain. When flying through a nimbostratus cloud one may observe that the cloud has a layered structure with thin layers of comparatively few water drops. In winter most of the cloud is in the form of ice crystals and snow flakes. The snow fall is often heavy and, in combination with big snow flakes, there is rapid accumulation on the ground in a northern climate.

Stratus

When raindrops from a nimbostratus cloud fall through the subcloud layer, water evaporates from the drops. The air in the lowest hundreds of metres gets more and more humid. If the drops from higher up are considerably colder than the subcloud air stratus clouds may form as a result of cooling and increasing humidity. Stratus clouds of this sort are also called scud (or fractostratus) and if the wind is strong they get a torn appearance. Close to the warm front they can cover the whole sky making it impossible to see the real nimbostratus cloud above.

Cumulonimbus

During the summer most of the rain falls in the form of showers. The cloud capable of producing heavy showers is called *Cumulonimbus*. Cumulonimbus (Cb) develops out of cumulus clouds (see colour plates 'Cb-Clouds I–IV' and 'The Ten Main Cloud Types'). If the atmosphere is conditionally unstable up to great heights the cumulus cloud can go on growing upward until it extends through the whole troposphere. The upper parts of the cloud reach temperatures that even during the summer are as low as -20 to $-30°C$. The upper part then consists of ice crystals, while the lower parts of the cloud are made up of water drops. This is what gives the cumulonimbus cloud its characteristic anvil-looking top and the fully developed cumulonimbus cloud is called *cumulonimbus incus* (incus meaning 'anvil'). When the upper parts of a growing cumulus congestus freezes it usually starts to rain from the cloud. The rain from a cumulonimbus cloud comes down quickly and violently and is usually preceded by strong, chilly wind gusts which at times can be very strong, easily reaching gale force. A storm of this kind can result in more than 20 mm of rain in less than an hour. The small town Curtea de Arges in Romania once had 206 mm of rain in twenty minutes! The cumulonimbus cloud is most common in the afternoon during the summer in northern latitudes, but occurs all year round in the tropical and subtropical regions. Similar clouds of less vertical extent also form during the winter giving snow showers. This is especially true if cold air during the winter flows over a warm water surface (such as the sea) picking up moisture at the same time. The southern shores of the Great Lakes in the U.S.A. receive several metres of snow each year from snow showers released by the relatively warm water in the lakes.

Over the warm tropical seas rain can fall from cumulonimbus clouds without the clouds bearing the shape of an anvil. The cumulonimbus is then called *cumulonimbus calvus* ('calvus' meaning 'bare').

70

Standing right under a cumulonimbus cloud gives the impression that the cloud is covering the whole sky but the size of a cumulonimbus cloud is fairly limited, ranging from a couple of square kilometres to 100–200 square kilometres. Often a cumulonimbus cloud is built up from several active convection cells from which the rain is falling. Many cells can congregate together forming rather extensive areas of cumulonimbus clouds. Such clusters are very common in the tropics.

Thunderstorms

The cumulonimbus cloud and its attendant weather are probably the most violent weather phenomena in middle and northern latitudes. In a well-developed cumulonimbus cloud there are very strong vertical movements with velocities of the order 20–30 m/sec. These strong up- and down-drafts create favourable conditions for the formation of thunder and hail.

Thunderstorms and lightning occur on the average five to fifteen days a year at places in Britain while in many areas of the U.S.A. thunderstorms are much more frequent; in the Midwest, for example, there are about fifty thunderstorm days a year. On the other hand, the West Coast has less than ten thunderstorms a year. In the inland region of Florida every day out of three is a thunderstorm day. The number of days with thunderstorms in Europe is shown in Fig. 20. Most of the thunderstorms occur during the summer months June, July and August.

Most thunderstorms develop inland and in hilly and mountainous terrain. A thunderstorm can be a beautiful and fascinating spectacle, but also causes considerable damage, mainly through lightning setting fire to trees, houses and barns, etc. Cattle and sheep may be killed and every year people are killed by flashes of lightning. An American investigation shows that more than half of the people killed by lightning are engaged in outdoor recreational activities, especially golf. Lightning kills more people on average than hurricanes or tornadoes, which are far less frequent.

What is thunder and what makes it develop? Unfortunately we don't yet know exactly how thunderstorms come about. But we do know that active cumulonimbus clouds with their strong vertical

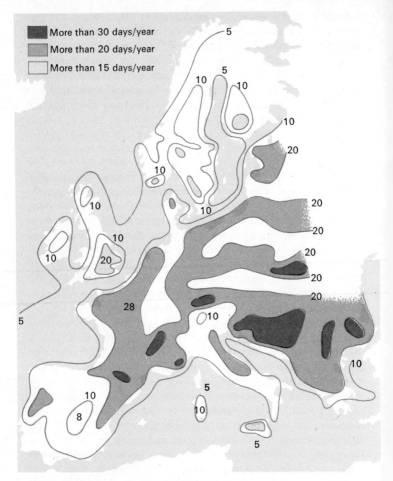

Fig. 20 Number of thunderstorm days in Europe.
The map shows the normal, annual number of days with thunderstorms in Europe. The greatest number of thunderstorm days is found around the Mediterranean with more than 30 days of thunderstorms in the highlands. In Great Britain most of the thunderstorms occur in the eastern and southern parts of England with generally more than 15 days each summer. Local maxima with more than 20 days a year are found in the Northampton area and to the east of Lincoln Edge. Ireland has less than 10 days a year of thunder while there is a local minimum with less than 6 days a year in southern Scotland. The west coast has generally less than 6 days a year. In north-western Europe more than 90% of the thunderstorms occur during summer.

motions and the simultaneous existence of supercooled water drops and ice crystals are capable of separating positive electric charges from negative ones in different regions of the cloud. Exactly how this separation of charges takes place is not yet known. The differently charged regions in the cloud produces enormous differences in electric potential which are necessary for lightning flashes. The potential in the lower parts of a thundercloud relative to the ground can be of the order of 300 million volts. The normal lightning flash that we usually say 'strikes down' does not really strike down but goes from a positively charged region in the ground to a negatively charged zone in the cloud. A flash of lightning is really nothing but a very strong electric arc. Actually we can produce arcs ourselves by, for example, rubbing a comb against a piece of fur. The comb gets charged and if we move it towards a piece of metal of some kind a small arc appears. During the winter when the air is dry, rubbing against the car seat might produce sparks when one touches a person or the side of the car, especially if one is wearing clothes made of synthetic fibres. Lightning is an arc millions of times stronger than our 'home made' arcs (see colour plate 'Cb-Clouds III–IV').

The lightning strike is rather complicated and is built up from several discharges of electricity. Usually it all starts with a weak discharge trying to find a smooth and easy way for the main discharge between the ground and the cloud. This first 'feeler', or 'stepped leader' as it is referred to, creates a lightning channel through which the main discharge can take place. The main lightning discharge lasts for about $\frac{1}{100}$ of a second. The first main discharge can often be followed by secondary, less energetic, discharges. Different lightning flashes contain different amounts of energy. Whether a lightning strike is going to cause a fire or cause mechanical destruction depends on its duration. The 'cold' lightning only lasts $\frac{1}{1000}$ of a second while the 'warm' energetic lightning can last for $\frac{1}{5}$ of a second.

The characteristic sound of thunder heard in connection with lightning is generated when the air along the lightning channel is explosively heated to a temperature around 30,000°C. This gives rise to a rapid change in air pressure along the channel which travels away from the lightning like a shock wave changing to sound waves. The powerful thunder can be heard miles away from the thunderstorm. The pitch of the thunder mainly depends on how energetic the lightning is. The more powerful the strike the lower the pitch of the resulting thunder. As the illustration in colour plate 'Cb-Clouds III–IV' shows, the lightning has a segmented structure. Each minute segment acts like a sound generator. Even if the lightning strike only takes about $\frac{1}{10}$ of a second it takes different times for the sound from the different segments to reach the listener. The speed of sound is 300 m/sec. and a lightning channel can be 5 km long. If the lightning is broadside on to the listener most of the sound energy reaches the listener at about the same time and the result will be heard as an intense clap of thunder. If the lightning on the other hand is oriented end-on to the observer the result will be heard as a protracted rumble or roll of thunder. Of course the sound is also affected by reflections from different objects, by the winds, and by the vertical temperature variations.

Distance to a thunderstorm

The speed of light is about one million times greater than the speed of sound. That means that a lightning flash is seen almost exactly at the time it occurs. Since the speed of sound is 300 m/sec., so that sound travels about 1 km in three seconds, we can work out the distance from the lightning by counting the number of seconds between the flash and the first moment we hear the thunder; dividing by three gives the distance in km.

Lightning does not only strike between the ground and the thundercloud, but also between different parts of the cloud. In tropical regions where the base of a thundercloud is usually quite high most of the flashes occur between different parts of the cloud. The lightning seeks the easiest way to the ground. The easiest way is often the shortest way, though there are exceptions to this rule. However, generally the lightning seeks its way to high objects like trees, or good electrical conductors such as metal objects.

If caught outdoors by a thunderstorm you should stay clear of high trees, telephone poles, flag poles and wire fences. You should also try to avoid the edge of a forest with high trees. The very best protection from lightning is in a car with a metal body and roof. An open car is no protection at all. A metal car acts as an insulator like a 'Faraday cage'; if hit by a lightning strike the current will follow the sides of the car down into the ground without penetrating inside the car.

When seeking shelter indoors it is wise to stay away from electrical appliances, the telephone and old-fashioned metal stoves. In bad storms, radio and television sets should be turned off and the aerials should be earthed. Big metal objects such as lifting cranes are definitely to be avoided during thunderstorms if they are not protected. It is also recommended to avoid cycling, riding, swimming and motor cycling.

If you are caught outside away from a car or house in which you can take shelter, experts recommend the following:
Avoid contact with metal objects such as an umbrella. Kneel down with feet and knees pressed together. Put your hands on your knees and bend forward. To lie down in a spread-out position increases the risks of getting hit by the lightning because your exposed area increases. Ditches and hollows are usually moister than the surrounding land and should be avoided. Also avoid hills and ridges; and especially avoid isolated trees.

The best thing to do when engaged in outdoor activities is to observe the development of the cloud sky. It is frequently possible to detect an approaching thunderstorm and seek shelter in good time.

Hail

Hail is normally not dangerous to man but every year causes widespread damage to crops in high risk areas, such as the Midwest in the U.S.A. and in parts of the U.S.S.R. In the U.S.A. the annual damage caused by hail is estimated to amount to $0·3 billion. Hail is a rough lump of ice, often called a hailstone, which can be several centimetres in diameter. In Coffeyville in the U.S.A. a hailstone has been found with a diameter of 13 cm, weighing 0·8 kg. The formation of hail requires a cumulonimbus cloud with extremely strong vertical velocities and the presence of supercooled water drops. Because of this, hail is rarely observed in the polar regions where clouds are made up exclusively of snow. Neither are they observed over the tropical oceans where the clouds normally are above 0°C and lack the ice crystals that can initiate freezing.

If a snowflake in a cumulonimbus cloud collides with a water drop, the drop will freeze on to the flake. When it has become larger the chances of further collisions increase and the smaller pellets adhere to the larger ones. When a small hailstone is hit by a supercooled water droplet two things can happen. Depending on the temperature of the small hailstone the drop can either have time to spread before freezing or can stick to the stone directly. In the upper colder parts of the cloud the latter usually takes place while in the lower warmer (but still below freezing) layers, the former happens. This gives the hailstone a layered structure. As shown in colour plate 'Cb-Clouds III–IV', a particular hailstone makes many trips up and down through the cloud before getting so heavy that the strong upcurrents can no longer hold it suspended in the air. The fact that hailstones can be as large as 10–15 cm shows that the vertical velocities inside a cumulonimbus cloud can reach values as high as 30–40 m/sec.

The presence of very strong vertical winds and supercooled drops causing icing on aircraft, hail and lightning in a cumulonimbus cloud makes them very dangerous to aircraft. Airliners are usually equipped with weather radar which makes it possible to detect cumulonimbus clouds and if possible avoid flying through them. Sometimes it is unavoidable, however, and if flying in such weather you are guaranteed an unpleasant experience.

Systems of cumulonimbus clouds

So far we have only looked at the cumulonimbus cloud as it grows out of a cumulus cloud in very unstable conditions. This is typical of the isolated shower or heat thunderstorm which is common in summer in warm air masses all over Europe. The variation over short distances in such a case is remarkable. In one place people can suffer an intense and violent thunderstorm, while 10 km away there may be bright and sunny summer weather. Cumulonimbus clouds can also occur in other circumstances, for example along a passing cold front. Sometimes thunderstorms can join together or form into long lines with intense cumulonimbus clouds. Such squall lines, which are very common in the U.S.A., especially during the spring, can be up to 500 km long. In Europe they usually form during the summer. In central Europe there seems to be a maximum in thunderstorm frequency at about 10 p.m. while in other places the maximum thunderstorm frequency is found in late afternoon. The Great Plains in the U.S.A. has a remarkably high frequency of night thunderstorms.

Make your own forecast of thunderstorms

If one wants to try to forecast the chance of thunderstorms one should study the cloud development in the morning. The presence of altocumulus castellanus is a serious warning that thunderstorms may form later in the day. A cover of high or middle high clouds of cirrus or altostratus has a suppressing influence on convection. After sunrise it usually takes some hours before the first cumulus clouds form. The later they come and the higher up they form the greater the chance for a fine day without cumulonimbus clouds. When cumulus has formed it is as well to check how it develops and grows. If the clouds remain small for several hours, then the air is too dry and stable to permit further growth. On the other hand if they start growing and form towering cumulus congestus, the risks are greater. The presence of a growth-preventing inversion aloft is evident if the growing cumulus spread out in the horizontal forming stratocumulus. In such a case we can exclude the possibilities of thunderstorms.

The growing cumulus congestus cloud is a thunderstorm signal and when we see the first fibrous tops of ice crystals we should know that under that cloud rain is falling. Thunder usually results when the tops grow high, up to 10,000–12,000 m. Listen for the thunder, and if the cloud is moving towards one it is best to look for shelter.

Precipitation

Nimbostratus is the common rain-bearing cloud, but in some cases ragged low clouds such as stratocumulus and stratus might also give drizzle and light snow. Stratocumulus is common in certain weather situations during autumn and winter when persistent easterly or south-easterly winds blow over northern Europe or along the northern parts of the Atlantic coast of North America. At temperatures above freezing point stratocumulus then often give drizzle, but at sub-zero degrees light but persistent snowfalls can occur. Stratus and stratocumulus also frequently give light rain or drizzle in the warm sector between a warm front and a cold front.

Nature exhibits a variety of different forms of precipitation. Drizzle consists of drops with diameters of up to half a millimetre. Rain and drizzle can be supercooled if they are falling from a supercooled cloud or from a warm cloud if the layer beneath the cloud is below freezing point. Freezing rain and drizzle are serious dangers to motorists. When the supercooled drops hit the surface of the road they immediately freeze giving very slippery road conditions. If the ground has been cooled for a longer period of time and is below $0°C$ warm drops may freeze when hitting the ground even if the air temperature is above freezing point.

Rain and drizzle can also freeze before hitting the ground. Frozen precipitation is different from snow, which forms by direct condensation to ice crystals. Granular snow consists of ice crystals on which rain or drizzle drops have adhered. Ice pellets are frozen raindrops while hail is the result of deposits of supercooled drops at different temperatures on snow flakes.

Sometimes precipitation can fall from different parts of the cloud at the same time. Snow and rain can occur mixed in the form of sleet.

During clear and very cold days in northerly latitudes light snowfalls can be observed from a cloudless sky. Water vapour in the air sublimates, i.e. is transformed directly from water vapour to ice, in the form of small ice needles that slowly fall to the ground.

Fog

Fog can be regarded as a special form of cloud, the base of which touches the ground. In a meteorological sense fog is defined as a layer of small water drops or ice crystals close to the ground reducing visibility to under 1000 m. Visibility between 1–10 km, caused by small water droplets, is called mist, in contrast to haze which is caused by the presence of small dust or smoke particles.

The drops in fog are very small with radii of about 0·001 mm. The liquid water content is also very small. At about —20°C the fog drops freeze and ice fog is formed. Between 0°C and —20°C we have freezing fog capable of giving rise to ice deposits in the form of hoarfrost. Fog or freezing fog is a serious danger to traffic. Despite all the advanced technical and electronic equipment used today by airliners and ships, fog is still a weather condition that can bring sea and air traffic to a virtual standstill. Fog makes it impossible for aircraft to land and many accidents have occurred at sea in dense fog. How the most common types of fog are formed is shown in the colour plate 'Fog'.

Other types of fog are *frontal fog* which occurs in the surface levels of a warm front and *orographic fog*. Frontal fog forms in a similar way to stratus clouds beneath a nimbostratus deck. Close to the front the air is thoroughly wetted by evaporation from rain drops and condensation can take place. Orographic fog is found when the cloud base is lower than the height of the ground over which the air is flowing.

Tornadoes

In the atmosphere the motion of the air often takes place in the form of whirls or eddies, ranging from the extended low pressure 'eddies' to the small eddies formed at street corners. In some cases the atmosphere can concentrate enormous amounts of energy into such movements. One of the most feared circular wind movements is the tornado. Tornadoes are found all over the world, but are most common in the middle latitudes of continents where cold and warm air masses meet. In the U.S.A. a tornado can develop devastating power and as a result many people are killed every year, especially in the southern, densely populated states.

The colour plate 'Tornadoes', shows the development of tornadoes and the destruction they can cause.

Tornadoes generally move quite rapidly at a speed varying between 40–80 km/h. In the initial stages, usually lasting for a couple of hours, air is sucked into the tornado and revolves at ever-increasing speed the closer it comes to the centre. However, when the tornado is in its mature stage the up-draft in the centre of the tornado turns into a strong down-draft of air.

A tornado does not usually occur alone, and many tornadoes can develop over a fairly large area. They are spawned by intense cumulonimbus clouds when the air is very unstable. A cumulonimbus cloud creates a circulation in the air, converging towards the up-draft region in the layer beneath the cloud. Sometimes this circulation can be concentrated into the whirling winds of a tornado. The giant cumulonimbus clouds in the U.S.A. are triggered off when warm and very moist air from the Gulf of Mexico moves northward towards the Midwestern states ahead of an advancing cold front approaching from the north-west. Very well-defined zones with large temperature contrasts between the cold air and the warm air from the south are created, especially during spring when tornadoes are most frequent. To the west of the cold front, temperatures can be below zero while the warm air is +25°C. In cases such as this conditions are favourable for the development of intense squall lines ahead of and at the cold front (see lower figures in the colour plate 'Tornadoes'). Violent thunderstorms, often with hail, accompany the squall line and feed the tornadoes.

The most devastating tornado catastrophe in the history of the U.S.A. happened on April 3–4, 1974 when more than eighty tornadoes hit the whole Midwest from the Gulf of Mexico in the south to Chicago in the north. During these two days 380 people were killed and 6,000 wounded. 13,500 homes were destroyed and the total damage amounted to $0.6 billion.

To meet the dangers that arise from violent thunderstorms and tornadoes, an effective warning system has been built up in the U.S.A. The reason why the catastrophe did not become greater than it actually did in 1974 was in many respects because of the accurate warnings. The main part of this warning system is the Severe Storm Forecast Center in Kansas City, where meteorologists try to estimate the risks for tornado development several days in advance. Local weather offices all over the U.S.A., equipped with weather radar, try to track every cumulonimbus cloud in its region of responsibility. Tornado watches and tornado warnings are issued through local television and radio programmes.

On May 11, 1970, the town of Lubbock in Texas was hit by intense thunderstorms, hail and a tornado. Twenty-six people died and thousands were made homeless. Below we can follow hour by hour the events in the town and how the warnings reached the public (WBO is an abbreviation for Weather Bureau Office).

Lubbock, May 11, 1970:

10.00 a.m. The Severe Local Storms (SELS) Unit amended the convection outlook issued earlier to include isolated thunderstorms with large hail expected in High Plains of West Texas east of Pecos late afternoon and early evening.

6.55 p.m. The Lubbock radar detected a moderate thunderstorm fifteen miles south of Lubbock Airport and about five miles south of the Lubbock City limits near the community of Woodrow.

7.50 p.m. Severe Thunderstorm Warning Bulletin issued for Lubbock, Crosby, eastern Hale and Floyd Counties, using a quick action form. Emergency Action Notification Signal (EANS) requested. Civil Defence was given warning by telephone. The switchboard operator was requested to notify Mr Bill Payne, the Civil Defence Director.

7.59 p.m. Radio station KFYO used EANS to alert other radio and television stations that a special warning message would be coming over the network. No 'commercials' given from KFYO from this time until 7.30 a.m., May 14.

8.10 p.m. Golf ball- to grapefruit-size hailstones reported by public to Lubbock WBO at 8.10 p.m. in the vicinity of Lubbock

Downs (about two to three miles south of city limits). Lubbock WBO checked with Amarillo Radar personnel on severe storm just south of Lubbock and found that cloud tops had increased to 55,000 feet (about 17,000 m).

8.10 p.m. Funnel cloud seven miles south of Lubbock Airport was reported by an off-duty policeman to Lubbock WBO.

8.15 p.m. Tornado Warning Bulletin issued for Lubbock, western Crosby, eastern Hale and Floyd Counties valid until 9.00 p.m. Quick warning form used. Additional note included that a hook formation was indicated on the Lubbock WBO WSR-1 radar at about the same time seven miles south of the Lubbock Airport, apparently moving north-eastward.

9.35 p.m. The Lubbock WBO WSR-1 radar indicated a tornado about seven miles south-west of the Lubbock Airport vicinity of 19th and Brownfield Highway. This information was relayed by two-way radio and telephone to Civil Defence. Sirens were sounded at 9.35 p.m. (this was the time that a patrolman reported a funnel in this part of city).

9.49 p.m. The Lubbock WBO lost all communications.

9.55 p.m. The Lubbock WBO personnel abandoned the WBO to take cover from the approaching tornado.

11.30 p.m. The Lubbock WBO relayed information to EOC by two-way radio that the tornado warnings were officially cancelled for all areas as storms had decreased to moderate intensity in the Lorenzo area.

(Adapted from Natural Disaster Survey Report No. 70–1, U.S. Department of Commerce.)

Optical phenomena

When light passes through the atmosphere it can be refracted, scattered and reflected by air molecules, water drops, ice crystals and solid particles suspended in the air. This modification can take place in numerous ways and we have already briefly discussed why the sky is blue during the day, but red at sunset.

Mirages

The refraction of light in the air can give rise to mirages or optical illusions. People see objects that cannot normally be seen. Air has the property of 'bending' light rays in the same manner as a lens, but in a very weak way. The ability of air to change the direction of light rays is measured by its refractive index, that is slightly greater than unity. This index increases with increasing air density (in decreasing air temperature), so that cold air has a higher refractive index than warm air. The effect of the variations of the refractive index can only be seen for light rays travelling a long

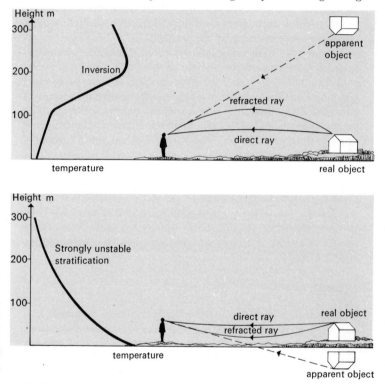

Fig. 21 Mirages.
The upper illustration shows the formation of mirages at stable stratification in which temperature increases strongly with height.
The lower illustration shows the situation on a hot day when the surface is very strongly heated, as in a desert. The direct rays show the normal way of light propagation while the refracted rays show the path of light when mirages occur. Mirages occurring under stable stratification make it possible to see objects lying below the geometrical horizon.

way through the air. Close to the ground that is only possible for light rays that are nearly parallel to the ground.

Mirages appear when we see objects that are below the geometrical horizon. They are observed when the temperature increases rapidly with height in the lowest parts of the atmosphere, i.e. the lowest 200 to 300 m. In such an inversion the refractive index decreases with height and the light rays bend down towards the surface of the earth (see Fig. 21). Mirages of this kind are common when warm air is streaming over a cold surface, such as snow or the cold sea during the spring.

Another type of mirage can be observed when the air temperature decreases strongly with height showing that density increases with height. The light rays bend upward. This type is most common during the summer when the air close to the ground is strongly heated, especially over roads, and slabs of stone, etc. The air is very unstable close to the ground and there is strong mixing of air in the vertical layers. The constantly changing air gives the layer above a road the appearance of a water-like, shimmering surface. In this case objects seem to be located below the horizon after reflection in the air.

This is the legendary mirage experienced by the person 'lost in the desert'. The blue light of the sky looks as if it is coming from the ground creating the impression of water pools and lakes. Oases appear on the horizon, but continuously evade the lonely, thirsty wanderer.

Rainbows

The rainbow is one of nature's most beautiful phenomena. It is seen during rain showers or fog when the sun is low. Rain drops have the same ability as a prism to split up the light into different colours (see colour plate 'Optical Phenomena'). When light falls on to a drop it is refracted at its entrance to and departure from the drop. When it hits the back surface of the drop the light is reflected, as shown in the illustration. In certain directions the light is enhanced while in others it is weakened. Different colours are refracted at differing angles. The spherical shape of the drop and the refractive index of water combine to produce a rainbow with a radius of 42° with red at the outer edge and blue/violet at the inner. Sometimes it is also possible to observe an outer, secondary and weaker rainbow with the colours reversed.

Haloes

Sunlight and moonlight can also be refracted by ice crystals, resulting in circular halo phenomena around the sun and the moon. The commonest one is the 22° halo which forms when light passes through cirrostratus clouds and ice fog. The 22° halo corresponds to the angle of minimum deviation for hexagonal prisms, which are typical in cirrostratus. The halo usually appears as a whitish, bright ring around the sun or the moon. If very strongly developed the inner edge may appear red in colour.

Mock suns

Another fairly common phenomena, especially in northern latitudes, is the mock sun, which is also a halo phenomena; the ice crystals in cirrostratus can take on different forms and orientation. When a cirrostratus cloud consists mostly of hexagonal plates, with their bases oriented downward, two mock suns appear, one on each side of the sun. The mock suns can become quite bright and, like the rainbow, show a range of colours.

Halo phenomena can become quite complicated, with a multitude of different haloes, mock suns, pillars, arcs and rings around the sun (and the moon).

6 AIR MASSES, FRONTS AND STORMS

Even if isolated showers seem to appear in a random and unpredictable way, the weather in the sea of air is related to fairly organized weather systems which also move and develop in a fairly regular way. With a knowledge of how the weather machine works and what the underlying physical processes are, it then becomes possible to make forecasts.

The most important weather system in the middle latitudes, i.e. 40°–70°N, is the polar front cyclone. In Britain the polar front cyclone is often called a depression, or simply a low, while in the U.S.A. people usually talk about storms. The polar front cyclone develops and moves in the west wind belt of the middle latitudes. The general west winds are, however, considerably disturbed by such features as high mountain ranges and the uneven distribution of land and sea. The turbulence of the air and the condensation of water vapour to water drops are other contributing factors, all of which make the prediction of storms not as easy a matter as one would think.

The first comprehensive and fundamentally correct theory of the mid-latitude depressions was put forward at the beginning of the 1920s. A group of distinguished meteorologists, both theoretical physicists and practical meteorologists, had been working in Bergen in Norway for several years to bring order and understanding to the mass of observations of the atmosphere which had already been accumulated. Outstanding scientists were, among others, the Norwegians Wilhelm and Jack Bjerknes, Sverre Pettersen, the Swede Tor Bergeron and Erik Palmén from Finland, and the picture we have today of the polar front cyclone is essentially the same as that presented by these men at the beginning of the 1920s.

These new ideas were gradually assimilated by other meteorologists so that by the beginning of World War II the ideas of the so-called Bergen school were generally accepted. The theories of these men were founded on the observations that are made every three hours all over the world of meteorological elements such as air pressure, temperature, wind, clouds and weather. These observations are made simultaneously and are called *synoptic observations*, the whole branch of meteorology that is concerned with the acquisition of these observations being called synoptic meteorology (see Chapter 7).

The Bergen school also introduced a series of new concepts in meteorology such as air masses, fronts and occlusions. These take into account that the driving force of the polar front cyclone is the temperature contrast between warm and cold air masses. In southern latitudes, in the tropical and the subtropical regions of the earth, on the other hand, it is the latent heat of water vapour in the air that can provide the energy for the hurricane force winds in tropical cyclones or help to form the widespread areas of precipitation in the so-called easterly waves in the trade wind belt.

Air masses

Over fairly large areas of the earth the air can acquire uniform properties in regard to temperature, humidity and vertical stability. Such a large volume of air is called an air mass. An air mass can be formed in a high pressure region where the air slowly rotates around the centre and gradually acquires the properties corresponding to the underlying surface. The Azores High (see colour plate 'Climate V–VI') is one source region for what is called tropical air.

Cold air masses form over the northern parts of the U.S.S.R. (particularly Siberia) and over northern Canada. The temperature of the air in an air mass is determined by the surface conditions and the radiation properties of the source region. The humidity on the other hand is largely dependent on whether the air mass forms over sea or land. There is an abundant number of classifications and subdivisions of air masses in meteorological literature. In its simplest form it designates the following air masses:

Tropical air (abbreviated T)
Polar air (P)
Arctic air (A)

An air mass can be of maritime origin and is then, in general, quite humid. Air that has spent a long time over the continent is on the other hand dry, and it may contain dust or air pollutants making it hazy. To differ between the two types of air, maritime and continental, we put a prefix m or c before the main air mass type. Thus, mP stands for maritime polar air while cT designates continental tropical air.

The source regions for different types of air masses are shown in the colour plate 'Air Masses' where typical temperatures and moisture content have been indicated. This illustration describes the mean conditions over the year. The southern and northern boundaries for the different air mass sources move towards the north during the summer and towards the south during the winter.

Maritime tropical air is usually created in the oceanic subtropical highs. The tropical air mass that forms over North Africa, the Arabian Peninsula and the Near East is hot and dry during the summer with temperatures up to $+45°C$. Dust and other minute solid particles, such as sand, make the air hazy.

Arctic air

The arctic air is very cold, dry and exceptionally clean, the cleanest of all air masses. As a result of this the visibility in arctic air is extremely good.

Polar air

Polar air is a sort of transitional air mass between arctic air and tropical air. The maritime polar air that very often invades northwestern Europe is formed over the North Atlantic when arctic air streams towards the south becoming heated over the warmer waters of the Gulf Stream.

When an air mass starts to move out from its source region it can either be warmer or colder than the surface it flows over. For example when tropical air streams towards the north during the winter it gradually encounters colder and colder ground or water surfaces. The air mass is therefore cooled in its lower layers. Since the moisture content in tropical air is quite high this cooling can easily produce fog, low clouds and drizzle. This is typical of a tropical air mass being displaced towards the north during winter. On the other hand if polar air or arctic air flows towards the south over, for example, the warm Norwegian Sea the air is considerably colder than the surface, sometimes the difference being as large as 10°C in the lowest 20 m. The air becomes unstable with resulting gusty winds, cumulus clouds and showers. Visibility is good because of the strong vertical mixing and broken clouds. The same air mass can appear as either a 'warm air mass' or a 'cold air mass' depending on how the temperature of the underlying surface varies.

Fronts

Temperature and humidity can differ substantially between different air masses and the differences frequently extend through the whole troposphere. The temperature difference between tropical air and polar air can be between 10°C to 20°C and an equally big temperature difference can be present between polar and arctic air. At higher levels the temperature differences can be even greater. In the colour plate 'Air Masses' the transitional zones between the air masses have been made rather broad and diffuse to show that their geographical location can vary considerably. Actually, the transitional zones between two air masses can in practice be very narrow, perhaps only 100 km. Such a narrow transitional zone between two air masses is called a *front*, The Bergen school studied the movements of the fronts and the development of depressions and saw them as a continuous struggle between cold and warm air masses where the cold air was constantly trying to extend its boundaries towards the south at the expense of the warm air. With the fresh memories of World War I in their minds it was natural to the Bergen school of meteorologists to name the transitional zone, the 'fighting line', a front. Fig. 22 shows a cross section through a front, from north to the left to south to the right. The front extends for thousands of kilometres lengthwise.

The front slopes through the troposphere with the cold air

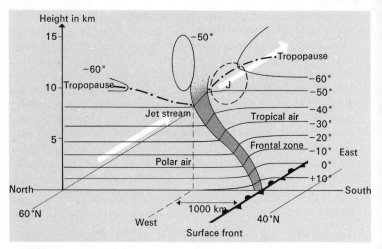

Fig. 22 The anatomy of a front.
The frontal zone, or the zone of transition between cold air in the north and warm air in the south, is shown in the middle of the illustration. The full lines are isotherms, i.e. lines indicating equal temperature. Taking one specific height as an example, such as 5 km, one can see that it is considerably colder to the north of the frontal zone than to the south (right). In the front itself, where the isotherms slope the most, temperature decreases the most from south to north. A certain temperature change is also taking place above the front in the warm air. The strongest winds are found at a height of 10 km, where they are concentrated into the jet stream.

wedged in under the warm air. The very abrupt temperature contrast between the air masses in a front and the slope is the result of an interplay between the rotation of the earth, the differential horizontal and vertical motion of the air at different latitudes and condensation and cloud formation. The rotation of the earth is a very important factor which is illustrated by the fact that fronts do not form at or close to the equator where the rotation around a vertical axis is zero or very small.

In association with fronts very strong winds usually occur in the upper troposphere. Since the front is characterized by very large temperature changes over a short distance the wind must increase rapidly with height in these regions. This leads to the creation of a strong jet stream just south of the front beneath the tropopause. This jet is called the polar front jet stream and is shown on the upper right hand page in the colour plate 'Winds III–IV'.

Warm front and cold front

A front can be fairly stationary for a period of time but sooner or later it starts to move. When the front moves so that cold air replaces warm air, usually by cold air moving towards the south, the front is called a *cold front*. On the other hand when warm air replaces cold, often by a northward movement of the front, it is called a *warm front*. The intersection of the front with the surface of the earth is the 'surface front' or just the 'front', and when drawn on weather maps the warm front is represented by a full red line. A cold front is drawn as a full blue line. A stationary front is drawn as a mixed blue and red line. This notation is used in all the colour illustrations in this book.

The front between the polar air and the arctic air can be as sharp as the *polar front* between the tropical air and the polar air, and it is called the *arctic front*. The arctic front, however, only extends up to a height of 3–5 km.

The slope of a front also varies considerably, but on average the cold front has a slope of 1/90 while the warm front has a slope of 1/150. That means that the fronts at 10 km height are some 900 m and 1500 m ahead of the surface front. The cold front has a steeper slope than the warm front, but even taking this into consideration the slope of the fronts have been substantially exaggerated in the illustration.

Polar front cyclones

A polar front cyclone generally forms and develops along a quasi-stationary polar front along which very warm air has been brought in close contact with cold polar air from the north. The greater the temperature difference the more powerful a storm we can expect. In the colour plate 'Polar front systems I–II' we can follow the growth of a polar front cyclone from a harmless little wave on an almost stationary polar front to a fully fledged storm. In the series of illustrations we imagine that we see the low pressure, the weather and the fronts from above as on a map. We further assume that we follow the cyclone in its motion when it moves from the west towards the east, for example from the middle of the Atlantic to western Europe.

The occlusion

In the later stages of its life the depression becomes what is known as occluded. The cross section C to D in the lowest figure in colour plate 'Polar front systems I–II' shows the occluded front. The cold air from the west has overtaken the warm front surface and a single belt of rain or snow is the result of the combined front. In this case the cold air to the west is slightly warmer than the cold air to the east of the depression. The true tropical air has been lifted up into mid-air and is no longer present at ground level. The warm sector has disappeared at surface levels and the whole of the central parts of the cyclone have filled with polar air. During the summer the cold air from the west or north-west is usually colder and heavier than the cold air in the front of the polar cyclone. The fast-moving cold front then lifts both the warm air and the eastern cold air upward. The cold air from the west wedges in beneath the cold air in the east giving the occluded front the shape in Fig. 23. This type of occlusion, common

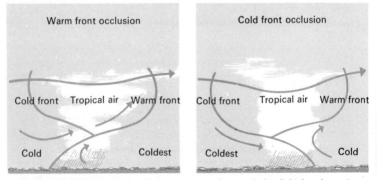

Fig. 23 A cold front occlusion (right) and a warm front occlusion (left) form in a polar front cyclone when the cold air from the rear of the low is colder or warmer respectively than the cold air ahead of the low. The clouds and the area of precipitation of the warm front occlusion are very similar to those of the warm front. The cold front occlusion on the other hand is very similar to the ordinary cold front with heavy, but rapidly passing, rain. During summer cold front occlusions often change into ordinary cold fronts associated with rain showers or thunderstorms.

during the summer, is called a *cold front occlusion* in contrast to the *warm front occlusion*, shown to the left, typical of winter time Atlantic depressions. The weather of the cold front occlusion resembles the true cold front with a narrow belt of rain associated with the front, sometimes accompanied by cumulonimbus clouds, showers and thunderstorms. When reaching continental Europe the cold front occlusion frequently changes into a simple cold front.

Colour plate 'Polar front systems III–IV' shows a bird's eye view of the mature and well-developed polar front cyclone in Fig. 4 of the previous plate. This plate shows in somewhat more detail the characteristic cloud and weather distribution in the polar front cyclone seen from the south.

Winds

The polar front cyclone is a huge wind system affecting an area as large as Europe. When a polar front cyclone is approaching from the west, air pressure starts to drop and at the same time the winds increase from a southerly direction. During the winter very cold air masses to the east can make quite a resistance to the approaching warm tropical air and gales from the south can occur. The closer to the surface warm front we come the more the wind increases. When the front passes the wind veers towards south-west and pressure becomes steady or does not drop as rapidly as earlier. The wind decreases somewhat and blows fairly steadily in the warm sector. The approaching cold front with its showery rain stirs up the wind again and in connection with the passage of the cold front heavy gusts precede the turning of the wind to north-west or west. The strongest winds occur in the rear, western sector of the depression. In probably the severest gale for which records are available in Britain, a wind speed of 125 mph (55·9 m/sec.) was recorded at Costa Hill, Orkney, on January 31, 1953, in the rear of a deep depression.

Different air masses with different temperatures are involved in depressions, but the distinctions between tropical and polar air can be hard to detect sometimes by a ground observer. The temperature of the earth's surface, over which the air has travelled when it reaches Europe and the vertical mixing of air have a pronounced influence on the air temperature as we feel it. In the summer, for example, the continental land masses are strongly heated while the surrounding seas like the Atlantic Ocean, the North Sea and the Baltic are comparatively cool. A warm front from the sea in such a case, with its rain and bad weather, generally causes a temperature drop. If the skies clear in the warm sector on the other hand, temperature can quickly rise. The maritime tropical air during the summer, coming from the Mediterranean, for example, can be muggy and unpleasant for northerners. A cold front, on the other hand, from the west or north-west during the summer, usually causes a drop in temperature and if followed by a major cold air outbreak, a chilly spell.

In winter time the situation is reversed. The continental land masses of Europe are usually cold, and in mid-winter are often covered by snow. The surrounding seas are warmer than the continent, and so a cold front or a depression from the west usually brings with it fairly mild Atlantic air. In such a case the temperature rises behind an occluded front. On the other hand we get the coldest weather during the winter in connection with cold air and cold fronts arriving from the north or east. Local topography and other local features influence the temperature substantially, a process which will be discussed in more detail in Chapter 8.

Make your own forecast of depressions

In order to forecast the approach of a polar front cyclone with its stormy winds and bad weather we can look for some characteristic signs. The first clue is given by the appearance of mare's tails – the high *cirrus uncinus* clouds. However, not all cirrus clouds signal bad weather but they should start making us suspicious. If we have a barometer and notice falling pressure at the same time as the cirrus thickens and cirrostratus takes over the sky we can seriously consider taking an umbrella with us. The 22° halo around the sun makes it possible to distinguish between cirrostratus and other 'contaminants' in the sky. If the pressure continues to fall and altostratus clouds replace cirrostratus and the wind increases it is only a matter of hours before the first rain or snow starts falling. By using 'the wind to our backs' rule it is possible to locate the low, and figure out where in the polar front cyclone we are going to be. From such simple observations it is difficult, however, to tell how long it is going to rain, how strong winds will be and what the weather behind the depression is like. In winter it is difficult to observe the clouds and make the forecast because usually low clouds, not connected with the storm, obscure the higher clouds associated with the approaching weather system.

With access to a simple aneroid barometer it is possible to improve the forecast. In contrast to a general belief it is not pressure itself, but rather its changes that are of importance in determining the future development of weather. On a simple aneroid barometer 'fair', 'unsettled' and 'rain' are often indicated in connection to air pressure. This information can be misleading. We may have fair weather when the pressure is low and cloudy and rainy weather when the pressure is high. Most aneroid barometers are equipped by an adjustable needle, and by turning it to coincide with the actual pressure we can observe the change of pressure, for example three hours later.

If pressure is falling rapidly at, say, 6–15 mb per three hours, we can be pretty sure that an intense depression is approaching which also should be confirmed by the clouds. The rapid pressure drop indicates that the low is moving very quickly. The rain should not last very long but can be heavy at times. It is usually followed by strong gusty winds and a rapid sky clearance, but showers might occur later. Air pressure should then be rising at the same rate as it was falling. In a weather situation like this we can expect other lows to follow, but also intermediate periods of clear skies and windy conditions.

If, on the other hand, pressure is falling slowly at 1–3 mb per three hours, that is an indication of a gradual take-over by a more extensive low pressure area and we must expect a couple of days with unsettled weather. Slowly rising pressure over a longer period shows that the weather is going to be dominated by a high pressure for some time to come. The skies are going to be clear and sunny and warm weather prevail, at least in summer, for a couple of days ahead.

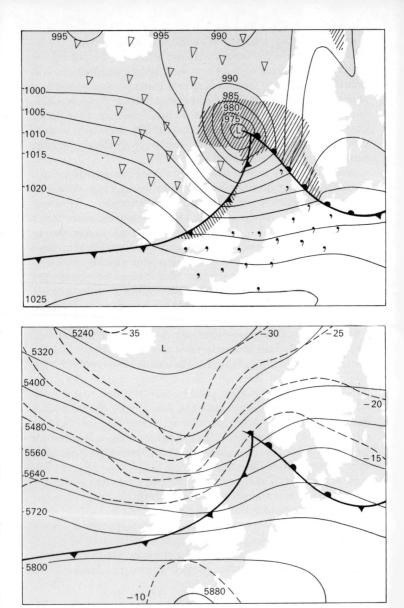

The air motion aloft

The rain and the clouds form in the 'free' atmosphere above the ground at roughly a height of 3–5 km. The weather and winds in a polar front cyclone affect the whole troposphere all the way up to the tropopause. A glimpse of the air motion aloft is given in the colour plate 'Polar front systems III–IV'. The jet stream curves around the surface low at a height of about 9 km, where we also find the highest cirrus clouds – the jet stream cirrus. Figure 24, however, shows both the pressure and fronts at the surface and the air flow at 5 km height. The example is of a severe storm that hit northern Europe September 22, 1969. The low at 5 km is located behind the surface low showing that the axis of the low is tilted towards the west or towards the colder air. In general the winds are much stronger at 5 km than at the surface in accordance with what we have said about the increase of the wind with height. At the beginning of the development of a new polar front cyclone no closed circulation, i.e. no closed isobar, is observed at 5 km, but only a trough behind the surface low. When the depression starts occluding the 5 km low is getting into a position right above the surface low. In the end phase, the dying stage of the depression, the low has a vertical symmetry right through the troposphere.

Rossby waves

While a depression has a lifetime of a few days the flow aloft on a larger scale can persist, and change only a little for periods of up to a week. The air in the free atmosphere above the ground moves in a wavy pattern that encircles the northern hemisphere and the largest wind speeds are concentrated to the jet streams at an altitude of 10 km. Colour plate 'Winds III–IV' shows the average

Fig. 24   The storm of September 22, 1969.
The upper chart shows the low pressure at the surface with isobars, fronts and typical distribution of weather at midnight (see also weather symbols in Fig. 29). The strongest winds are found on the western side of the low where the isobars are closest together. The lower chart shows the air flow at a height of 5 km. The full lines correspond to the isobars on a surface map while the broken lines are isotherms – lines indicating equal temperature.

position of the jet stream in winter. The mean jet stream moves towards the south over North America and over Central Asia. It also takes up a more southern position over the South Atlantic in connection with the subtropical jet stream. The flow at 5000 m is very similar to the flow at 10 km. The zigzag pattern with bulges towards the south and north is to a large extent due to the influences of the big mountain ranges and the distribution of land and sea over the northern hemisphere. The Rocky Mountains and the Himalayas, with help from the heating the air experiences over the oceans, fix the positions of the fluctuations. If we take a look at the 500 mb map corresponding to the flow at a height of 5 km over the northern hemisphere in colour plate 'The General Circulation III–IV' we can see that on a specific day the wave pattern is considerably more complicated than the map for the mean conditions shows. The picture is complicated by the presence of several polar front cyclones distorting the upper waves to fit their own cyclonic circulations. Despite this it is still possible to observe the two major southward bulges in the wavy pattern of the west wind belt, one over eastern U.S.A. and another one over east Siberia. If one carefully studies the positions of the jet stream and the number of zigzags or waves around the globe, significant variations of these over periods of a couple of weeks are evident. Another rhythm of three to four weeks is also present. Sometimes, especially in winter, it is only possible to discern two or three waves around the northern hemisphere and such a situation is usually accompanied by very strong jet streams with wind speeds up to 200 knots. Migrating lows and highs move quite rapidly from the west towards the east in the strong westerlies. After a week or so it looks like the whole pattern is starting to buckle. More waves form along the jet stream and the number of zigzags increases to four or five. The average west wind decreases and an alternative pattern is formed with southerly and northerly winds where the jet stream either pushes up to the north or down to the south. Such a weather situation is characterized by warm air moving far up northwards while in other places cold air is transferred to the subtropical regions. Such a pattern can develop during the spring over the eastern Atlantic Ocean and the eastern Pacific. A major portion of warm air can be stranded at northern latitudes in the centre of a massive upper air high pressure forcing the jet stream to split in front of the high, one branch turning towards the north, another veering deeply towards the south. At the same time a big pool of cold air has been cut off far to the south where it slowly becomes warmer. Such an air flow is generally called a *blocking* situation and can persist for weeks. During spring, when the land in Europe starts to heat up such a blocking frequently forms over England moving slowly towards Scandinavia, giving dry and pleasant spring weather during May. When, after a week or so, the blocking high pressure zone retreats out over the Atlantic Ocean one of these nasty backlashes of weather in the form of a massive cold air outbreak occurs over north-western Europe, coming as a shock for people who thought that the winter was definitely over.

Variations in the position of the jet stream and the movement of the long waves in the upper troposphere thus greatly influence the weather at the surface. A depression on the polar front is often only one in a whole family of polar front cyclones moving like ripples below the large-scale zigzag pattern formed by the jet streams. This is also evident from the satellite picture in the colour plate 'The General Circulation III–IV'. In one wave, also called a Rossby wave after the Swedish-American meteorologist Carl-Gustav Rossby, several disturbances along the polar front can be present at the same time in different stages of development. While an individual cyclone can move with a speed of 40–80 km/hour, the Rossby waves move more slowly, their speed varying between 20 and 50 km/h. In other cases, as in connection with blockings, they can be stationary or even move towards the west. The Rossby waves 'steer' the shorter cyclone waves and determine where the depressions are going. The positions of the Rossby waves are not only determined by the mountain ranges and the land–sea distribution, but by more complicated interactions in the atmosphere.

Sometimes extensive polar front cyclones can get so vigorous that they seem able to change the size and position of the Rossby waves themselves. Often this is the prelude to a major change in the upper air pattern when the strong westerlies with two or three waves around the globe break down into four to six waves carrying warm air far to the north and cold air far to the south.

The periodic changes in the flow patterns in the westerlies and the possibility of predicting the motion and development of Rossby waves is the foundation for forecasts of up to five days or longer. Forecasts of this length have been facilitated by the development of fast and big electronic computers, but their accuracy is limited.

Where does the polar front cyclone get its energy from?

In a polar front cyclone the air is in motion over a very large area. In the vast wind system occupying the whole troposphere enormous amounts of energy are present. The energy, in the form of kinetic energy, amounts to 3 thousand billions kilowatt hours or one thousandth of the energy the earth receives from the sun every day. This energy is equivalent to the energy released in a thousand hydrogen bombs. Where does the polar front cyclone get its energy from? A simple explanation was put forward in the second decade of this century by the German meteorologist Margules. By studying Fig. 25 we can get an idea of how it works.

In Fig. 25 (left), we imagine in a simplified way a large volume of air enclosing a polar front. The polar front slopes towards the north with the cold air acting like a wedge under the warm air. The cold air has greater density than the warm air and is therefore heavier. When the cyclone starts to develop and the fronts get occluded the end result will be as schematically shown in the right-hand diagram. The warm air is floating entirely on top

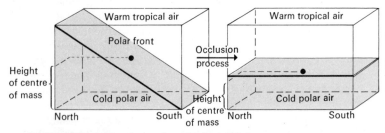

Fig. 25 The generation of energy in a polar front cyclone.
When the low develops, warm air is pushed upwards by the cold air which is sinking down beneath the warm air. The net effect is a lowering of the centre of mass by means of which potential energy is transformed into kinetic energy, i.e. the winds increase around the low, as shown in the diagram.

of the cold air which is at the bottom of the box. In the beginning the centre of gravity is higher up than in the end phase. The net result of the storm development is that the centre of gravity is lowered which in principle is similar to dropping a stone from one height to a lower height when the potential energy of the stone decreases and the stone picks up speed, i.e. kinetic energy. This is what happens in a polar front cyclone. The winds increase around the low; when the occlusion process has lifted all the warm air up from the ground and the low is filled with cold air in its lower layers the centre of gravity has reached its lowest position and all potential energy is consumed and converted to motion. The ground friction slowly dissipates the kinetic energy and the low pressure area fills up.

The warmer the warm air and the colder the cold air, i.e. the greater the temperature contrast between the tropical and polar air, the more potential energy is available to the polar front cyclone and the more intense a depression is to be expected. Clouds and precipitation can also be of significant importance in the development of the storm. Condensation in the widespread cloud masses in a low in connection with the warm front releases large amounts of latent heat that contribute to heating the already warm air. The cyclone gets an extra injection of potential energy. It is also commonly observed that intense polar front cyclones contain much warm and humid air, tropical air and widespread precipitation.

Cyclone tracks

Most polar front cyclones form over regions where the temperature contrast between warm and cold air masses is at its greatest. This is true, for example, along the east coast of North America. Warm and moist air streams northward clashing with very cold air from the inner parts of eastern Canada. The ocean areas around Iceland are another birthplace for depressions and many of the cyclones reaching north-western Europe form in this region.

During the summer the continent is strongly heated while the Atlantic stays comparatively cool. In the border zone between the

Atlantic, cool air and the continental warm air lows can deepen and develop moving from south Europe towards central or north Europe.

The polar front cyclones are generally weaker during the summer than during the winter and take a more northerly route. During winter and especially during January and February the polar front is displaced far to the south and depressions move across the Mediterranean. Cold arctic air masses then generally cover northern Europe. Most storms hit western Europe during the autumn when the temperature contrasts between polar and tropical air are greatest. Different typical depression tracks and weather situations are shown in the colour plates 'Polar front systems V–VIII' during winter and summer. These Plates should make it easier to follow and understand the daily forecasts

The possibilities for meteorologists to forecast the weather are made more difficult by the fact that not all fronts and depressions behave as regularly as expected. Frontal disturbances may in many cases not deepen and develop into vigorous storms. The meteorologists speak in such cases of 'stable waves'. But even stable waves may give a lot of precipitation, especially during summer when the tropical air is very moist.

Most of the polar front depressions which have formed out over the North Atlantic have passed through their most active stages before they reach north-west Europe. There are, however, notable exceptions, and winds in excess of 100 mph (44·7 m/sec.) are not uncommon on the western and northern coasts of the British Isles.

The storm which caused the most severe damage and loss of life in recent years, was the one on January 31/February 1, in 1953, when a slow moving, very deep depression was between North Scotland and Denmark, with the result that extremely strong northerly winds were blowing down across the North Sea.

These severe gale winds occurred at the time of a high tide (fortunately not a spring tide) in the southern North Sea, so that there was an additional rise in sea level of some 9–11 feet (2–3 m) which over-ran the sea defences of eastern England and the Netherlands. There was very serious flooding of salt sea water across low-lying areas, and considerable loss of life, especially in the Netherlands.

A comparable disaster occurred some 200 years previously. The severe gales caused extensive damage in the forests of Scotland and northern England, but it was the coincidence in time of two natural threats, the gales and the high tides, which brought about the most serious consequences.

Tropical cyclones

Tropical cyclones are intense and violent circular storms that form out over the tropical oceans and can cause severe damage in coastal areas around southern U.S.A., the islands in the Caribbean Sea, Japan and the coasts of the Far East. They are called Hurricanes in the U.S.A., Typhoons in Japan and the North Pacific, and Cyclones in Australia and the Indian Ocean. A remarkable feature with these violent weather systems is that they only form where the temperature of the ocean surface exceeds $+27°C$. The tropical cyclone is the most devastating of all weather phenomena. It has been estimated that 80% of all people who have died in different storms have died in tropical cyclones, which is equivalent to some 5000 people a year. The worst storm in living memory swept across the low land of what is now Bangladesh and the delta land along the Bay of Bengal on November 12–13, 1970. Storm surges 10 m high associated with extremely strong winds drowned 300,000 people. Tropical cyclones cause great material damage every year, and are very seasonal in occurrence.

Hurricanes are followed by winds easily exceeding 60 knots, torrential rains causing flooding and storm surges drowning low-lying land along the coasts. In a strict meteorological sense the expression tropical cyclone is only used for a tropical weather system where the wind speed exceeds 32 m/sec. or 65 knots. Below this wind speed they talk about tropical storms. In rare cases the wind speed in a tropical cyclone can reach 150 knots or 80 m/sec.

New technical aids have recently made it possible to study in detail the movement and development of this extreme type of weather. Satellites and radar are now in use throughout the world. In the U.S.A., the Weather Bureau started quite early with special reconnaissance flights over the South Atlantic to detect and follow the development of the tropical cyclones at the earliest possible stage, so that it would be possible to warn about the approaching danger in good time. Nowadays the satellites have made possible a continuous monitoring of the vast ocean areas and no tropical cyclone can develop undetected.

Structure of tropical cyclones

Tropical cyclones differ in many respects from the polar front cyclone of the middle latitudes. Their horizontal dimensions are considerably smaller. A normal tropical cyclone is hardly larger than 500 km but the strong, hurricane force winds are concentrated into a ring with a radius of 150 km. Colour plate 'Tropical Cyclones' shows the structure of a tropical cyclone and how it forms and develops, for which process the rotation of the earth is essential. Figure 26 below shows that at the equator the earth

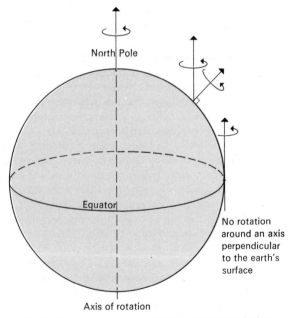

Fig. 26 The rotation around a vertical axis at different latitudes.

does not possess any rotational motion around an axis perpendicular to the surface of the earth. An air particle moving horizontally at the equator is unaffected by the earth's rotation. In the discussion of winds in the middle latitudes we saw that the rotation of the earth had a deciding impact on shaping the motion. Close to or at the equator the rotation of the earth has little or no effect on the motion and the wind. In the tropical cyclone, rotation plays an important role in organizing the whirling motion in such a way that hot and moist air can spiral in towards the centre of the cyclone where it is ascending, forming the intensive cumulonimbus clouds. Tropical cyclones cannot form close to or at the equator even if the ocean temperature is above $+27°C$.

In this connection we can also see another important difference between the polar front cyclone and the tropical cyclone. In the mid-latitude storms the source of energy was the temperature contrasts between warm and cold air which could be released in an extensive wind system. In the tropical cyclones the energy source is elsewhere, in the hot and very humid air over the warm, tropical oceans that can be liberated in the strong up-currents of the towering cumulonimbus clouds. The rotation of the earth provides the organizing mechanism for this and once it has started, the whirling motion of the cyclone itself sucks in more humid air towards the inner ring of violent rain storms and thunderstorms.

Over the oceans the cyclone can go on living for days and may travel across the whole of the South Atlantic before reaching the West Indies or U.S.A. If the tropical cyclone comes in over land it usually dies rapidly. The main reason for this is that the cyclone is cut off from its supply of warm and moist air, but a contributing factor is the increased friction over land that slows down the moving air.

Tropical weather systems

The tropical cyclones form and travel in the general east to north-easterly airflow in the trade winds north and south of the equator. They generally move from east to west even if their motion sometimes can be rather erratic. In this general easterly flow other tropical weather systems also form and move such as the so-called *easterly waves.* These are weak systems built up by clusters of cumulonimbus clouds, which are quite capable of giving large amounts of rain. Many tropical cyclones have originally started as a harmless easterly wave, but not all easterly waves develop into a tropical cyclone. Why that is so is not known exactly, partly because of the very limited number of meteorological observation stations over the tropical seas. This makes it very difficult to study the growth and development of such a small weather system as a tropical cyclone. In order to get a better understanding and knowledge of the weather and weather systems in the tropics and the formation and development of tropical cyclones in particular a very big international research programme was launched in the South Atlantic in August and September 1974. In this project, which was called GATE, 35 countries with 25 ships, 11 aeroplanes and 3500 technicians and scientists participated.

Although a tropical cyclone generally moves in the easterly winds of the trade winds, their movement sometimes can be quite erratic and capricious. A small error in the forecast of the motion of such a small, but intense, weather system can have disastrous consequences. In the countries which are frequently hit by tropical cyclones, a lot of money has been invested in warning and surveyance systems to make it possible to warn for approaching storms. Far reaching radars and geostationary satellites are parts of these systems.

Tropical storms can sometimes travel far enough north to become part of the prevailing westerlies of the higher latitudes, in which case they take on the properties of an intense polar depression.

For thousands of years man has observed 'the heavens' in an attempt to decipher tomorrow's weather.

Today he no longer stands in such awe of the forces of nature as their mysteries are gradually solved; 'portents' have been replaced by optical phenomena and observers throughout the world now use computers to assess the state of the air above. The millions of figures which are electronically produced are processed into forecasts for weather conditions several days ahead.

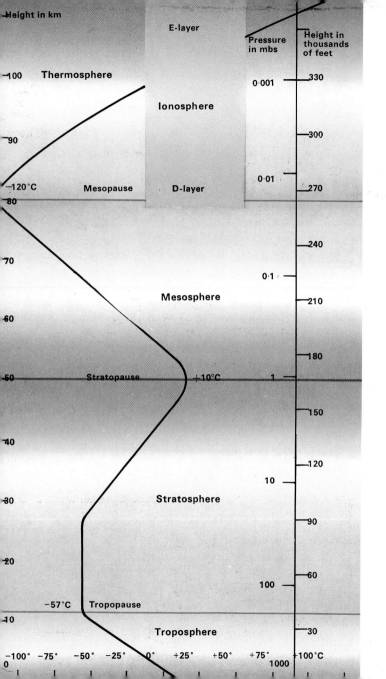

The Atmosphere I–II

Aurora Borealis

Height in km

Noctilucent Clouds

The atmosphere is the envelope of air surrounding the earth. The temperature varies considerably with height, so that the atmosphere is divided into different layers with different properties.

The diagram on the left shows the vertical variation of temperature in the atmosphere on the average. The average surface air pressure is 1013 mb and the average surface temperature is +15°C. At a height of 80 km the temperature reaches its lowest value, −120°C, in the summer, and here the pressure is just 1/100,000 of the pressure at the surface. At this altitude noctilucent clouds form which can be observed in northern latitudes during the summer. The electrically conductive Ionosphere is found at an even higher altitude. At this level Aurora Borealis forms when the air molecules are hit by highly energetic radiation from storms on the surface of the sun. Aurora Borealis is very rarely observed south of 50°N.

Meteorological Forecast Balloon

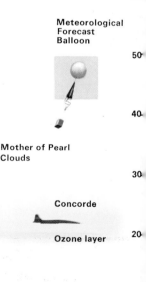

Mother of Pearl Clouds

Concorde

Ozone layer

What we normally call weather, such as rain, snow and thunderstorms, takes place and forms entirely in the Troposphere, the lowest ten km of the atmosphere. Mount Everest penetrates almost through the whole Troposphere. An ordinary jet liner flies at the limit between the Troposphere and the Stratosphere. The supersonic aircraft fly at about the same altitude as the ozone layer which protects us from the intense ultra-violet radiation from the sun.

Himalayas

100

90

80

70

60

50

40

30

20

10

0

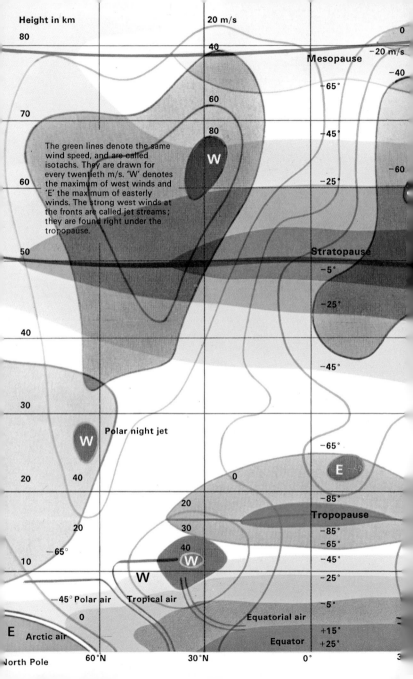

Height in km

80

70

60

50

40

30

20

10

20 m/s

The green lines denote the same wind speed, and are called isotachs. They are drawn for every twentieth m/s. 'W' denotes the maximum of west winds and 'E' the maximum of easterly winds. The strong west winds at the fronts are called jet streams; they are found right under the tropopause.

40

60

80

W

40

W Polar night jet

0

20

30

40

W

W

−65°

Tropical air

−45° Polar air

0

E Arctic air

North Pole

60°N

30°N

0°

0

−20 m/s

Mesopause

−40

−65°

−45°

−60

−25°

Stratopause

−5°

−25°

−45°

−65°

E

−85°

Tropopause

−85°

−65°

−45°

−25°

−5°

Equatorial air

+15°

Equator +25°

The Atmosphere III–IV

The most important properties of the atmosphere can be summarized in one illustration, an average cross-section from the North Pole to the South Pole for January (left). The large illustration shows a detailed and enlarged version (viewed sideways) of the area coloured pink in the small illustration below. Mean values of temperature and wind speed are shown by sets of lines. The lines separating different shades of orange, via yellow to blue, fields are isotherms – lines connecting points with the same temperature. They are numbered from +25°C to −125°C.

−125°
−105°
−85°
−65°
−45°
−25°

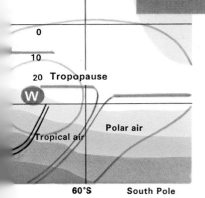

North Pole
80
40
0

W

E

E

Equator

South Pole

Double brown lines in the lower part of the illustration (left) show zones of transition from one air mass to another. These are called fronts, of which the two most important are the polar front and the arctic front which separate tropical air from polar air and polar air from arctic air. In January the northern hemisphere experiences winter and the southern hemisphere summer; there is no arctic air in the southern hemisphere. The heavy blue lines show the tropopause, which is highest over the equator.

0
10
20 Tropopause

W

Tropical air

Polar air

60°S   South Pole

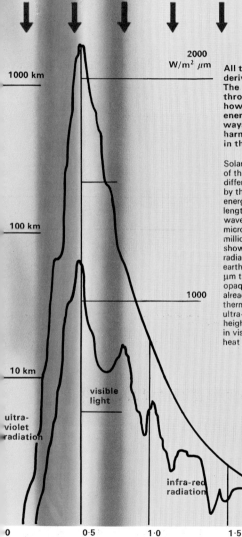

solar radiation

10,000 km

2000 W/m² μm

1000 km

100 km

10 km

1000

visible light

ultra-violet radiation

infra-red radiation

0   0·5   1·0   1·5   2·0   2·5 μm

All the energy in the atmosphere derives its origin from the sun. The way of solar radiation down through the atmosphere is, however, complicated and a lot of energy disappears in different ways on its descent. Radiation harmful to man is mostly absorbed in the upper atmosphere.

Solar radiation reaching the outer edge of the atmosphere is distributed on different wave-lengths. This is shown by the upper curve, which shows the energy content for different wave-lengths in watts/m². μm, where the wave-length is expressed in micrometres (μm) equal to one millionth of a metre. The lower curve shows how much remains when solar radiation reaches the surface of the earth. For wave-lengths less than 0·3 μm the atmosphere is practically opaque. Gamma- and X-rays have already been absorbed in the thermosphere and the major part of ultra-violet radiation is absorbed at a height of 40 km. Most energy is found in visible light and infra-red light, i.e. heat radiation.

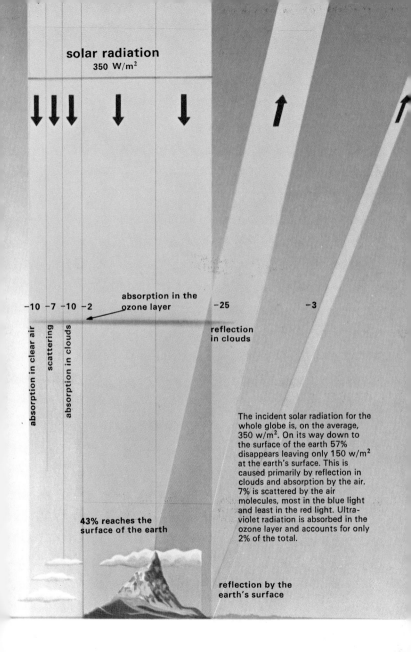

solar radiation
350 W/m²

absorption in the
ozone layer

−10 −7 −10 −2   −25   −3

reflection
in clouds

absorption in clear air

scattering

absorption in clouds

The incident solar radiation for the
whole globe is, on the average,
350 w/m². On its way down to
the surface of the earth 57%
disappears leaving only 150 w/m²
at the earth's surface. This is
caused primarily by reflection in
clouds and absorption by the air.
7% is scattered by the air
molecules, most in the blue light
and least in the red light. Ultra-
violet radiation is absorbed in the
ozone layer and accounts for only
2% of the total.

**43% reaches the
surface of the earth**

reflection by the
earth's surface

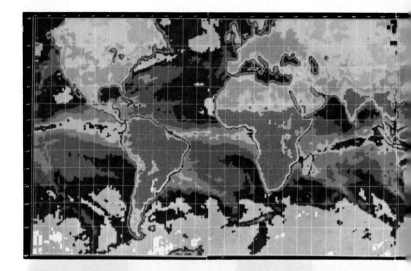

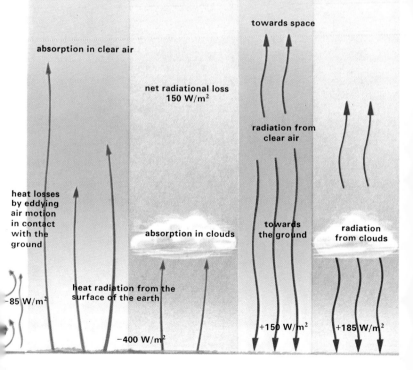

towards space

absorption in clear air

net radiational loss
150 W/m²

radiation from
clear air

heat losses
by eddying
air motion
in contact
with the
ground

absorption in clouds

towards
the ground

radiation
from clouds

−85 W/m²

heat radiation from the
surface of the earth

−400 W/m²

+150 W/m²

+185 W/m²

The Atmosphere VII–VIII

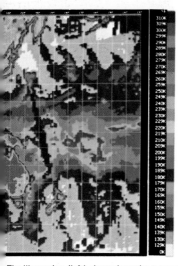

The earth's heat radiation as registered by the American Satellite NIMBUS , 12–16 January, 1973, expressed in the corresponding temperature.

Each day and night the earth receives 4 million billion kilowatt hours of solar energy. A similar amount must leave the earth so that the temperature of the earth does not change. Part of the solar radiation is immediately reflected back to space by the surface, but the main part is re-emitted as long wave, infra-red, radiation from the air and the earth's surface. The incoming and outgoing radiation is, however, unevenly distributed over the earth.

The illustration (left) shows how the temperature of the earth's surface is affected by long wave thermal radiation. Among other substances, the air contains water vapour and carbon-dioxide capable of absorbing heat radiation from the ground. Part of this radiation is re-emitted back towards the ground thus reducing the heat loss of the earth towards space and in fact this heat loss is considerably less than if no vapour and carbon dioxide had been present. This is usually referred to as the green-house effect. Cloud acts as an even more efficient lid on the atmosphere; if a cloud deck is forming on a clear cold night, temperature can rise by 10–20° in a couple of hours. Some heat also leaves the ground by the whirling, eddying motion of the air close to the ground.

The upper left illustration shows dramatically the amount of heat radiation leaving the earth as measured by the American satellite NIMBUS 5, 12–16 January 1973. Heat radiation has been converted to corresponding temperatures (see temperature-colour scale to the right). This is based on the fact that the warmer a surface is the more heat it radiates. The red areas in the tropical continents are the warmest. The blue ribbon around the equator is

the cold upper parts of the clouds of the tropical convergence zone. (See also *General Circulation I–IV*.)

The illustration below shows how the incoming and outgoing radiation vary from the pole to the equator. At the equator the sun is in its zenith and the incoming solar radiation (red arrows) is greater than the outgoing infra-red radiation (green arrows). At the pole the sun does not reach very high above the horizon and the situation is reversed. At about 40°N they are the same. As a consequence of this imbalance the tropics constantly receive more heat than they lose while in the polar regions there is a constant loss of heat. To prevent the equatorial regions from constantly getting hotter and the polar regions from constantly getting colder there is an unceasing transport of heat away from the equator towards the poles by means of the atmospheric wind systems and by the ocean currents.

W/m

−300

surplus

deficit

outgoing
thermal
radiation

incoming
solar
radiation

−200

−100

90°N   60°N   38°N   30°N   Equator

Optical Phenomena

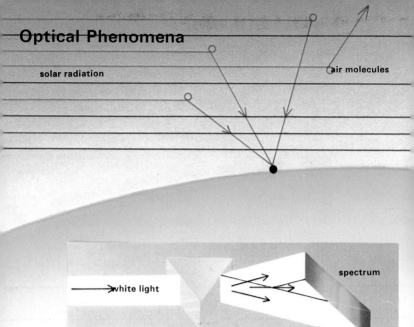

solar radiation

air molecules

→white light

prism

spectrum

Water drops, ice crystals and pollutants give rise to different optical phenomena such as rainbows and haloes. A halo is a coloured ring around the sun with a radius of 22° formed by cirrostratus clouds (See The Ten Main Cloud Types).

The blue colour of the sky (below right) is caused by the scattering of light by the air molecules (top illustration). Blue light is scattered more than the other colours and reaches the surface from all directions. When light travels a long distance through the atmosphere red light is scattered and the sky turns reddish. This is especially true at sunset (below left) and also occurs if the air is polluted.

The rainbow is one of the most beautiful phenomena of the atmosphere. It is visible in connection with rain or fog when the sun is relatively low and its formation is due to the refraction of sun light by rain drops. The rain drops act like a prism splitting the 'white' sun light up into different colours, because different colours are refracted differently (centre left). The diagram below shows how light is refracted and reflected in a rain drop. Red is refracted the most and blue the least; the rainbow is red on the outside and blue on the inside. The spherical shape of the drops and the refractive index of water makes the radius of the rainbow about 42° (above). Sometimes a weaker, outer secondary rainbow may also be observed, but with the order of colours reversed.

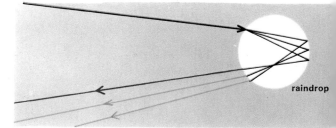

raindrop

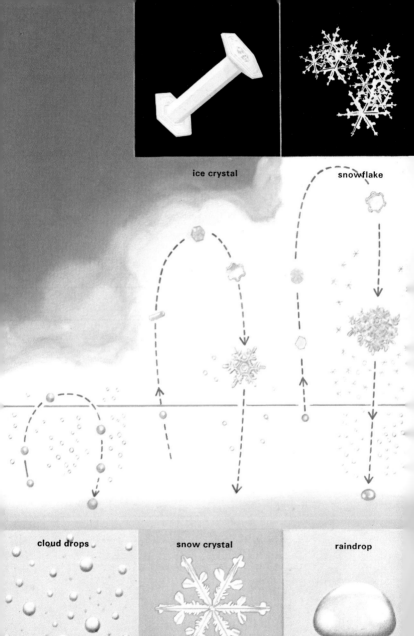

ice crystal

snowflake

cloud drops

snow crystal

raindrop

Cloud Physics

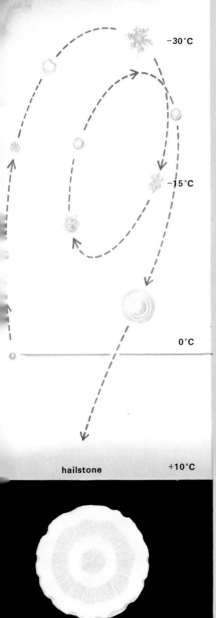

−30°C

−15°C

0°C

hailstone   +10°C

The relative sizes of different drops

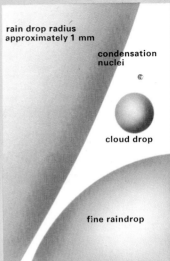

rain drop radius
approximately 1 mm

condensation
nuclei

cloud drop

fine raindrop

Clouds consist of minute water
drops which are formed when
rising air is cooled to saturation
point. Collisions occurring between
the drops suspended in the upgoing
air current can produce much larger
drops which eventually fall out of
the cloud as rain.

In thick clouds it is more common for
some cloud drops to freeze into ice
crystals which can rapidly grow into big
snow crystals by condensation. They
start falling, melt if they pass the freezing
level and collide with other drops or
snow crystals, finally reaching the
ground as rain. Rain drops are not
spherical but because of the air
resistance are 'roll' shaped.
Very strong up- and down-draughts
occur in thunderclouds and in these
vertical currents ice particles are carried
up and down several times. Usually
they grow larger by colliding with
supercooled water drops. In the lower,
warmer, parts of the cloud they usually
spread before freezing while in the upper,
colder, parts they immediately freeze.
Hail formed in this way gets an onion-
like appearance with a layered structure.

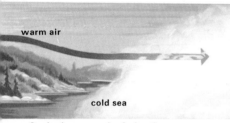

warm air

cold sea

Sea fog is an example of advection fog. This type of fog is very common along the coast of California and frequently rolls in over the Golden Gate Bridge in San Francisco. Advection fog forms when warm humid air flows over a colder surface. The air close to the surface gradually cools and the water vapour condenses to fog droplets (below).

Fog is simply a cloud touching the ground and consists of small water droplets or, at low temperatures, small ice crystals. Meteorologically, fog is usually defined by visibility being reduced below 1000 m. The three commonest types of fog are advection fog, radiation fog and steam fog. Fog can also form at surface fronts, when it is called frontal fog, and when warm moist air is forced upward over a mountain, called orographic fog. Fog can be a beautiful phenomenon, but is also a hindrance to air and road traffic and sea transport.

cold surface

mild humid air

Fog

Steam fog (right) is formed in the same way as steam from a pan with boiling water. The warm humid air rapidly mixes with the surrounding cold air.

steam fog

0°C   0°C
+10°C

Radiation fog (below left) usually forms during clear nights with weak winds. During the night the air closest to the ground cools because of long wave radiation from the surface (below). Radiation fog starts as thin patches of fog when temperature gets low enough to allow condensation to form. By continued cooling by radiation the fog then grows in depth and can become 100–200 m thick. Cold air has a tendency to flow down into hollows and valleys and because of this radiation fog usually starts forming in those places. Moist or marshy ground facilitates the formation of radiation fog. It is most common in early autumn when nights are growing longer, but the air is still fairly humid.

long wave radiation

+10°C

+5°C

Cb-Clouds I–II

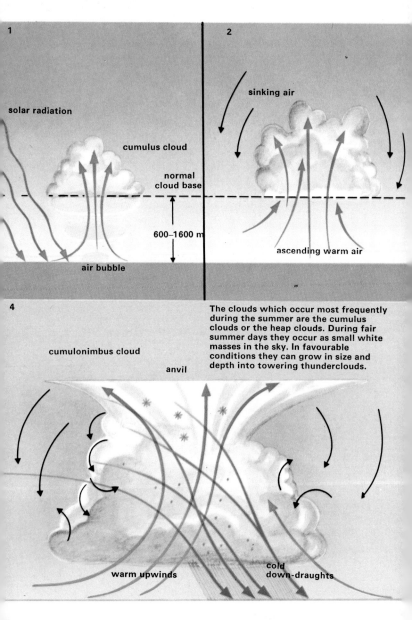

1

solar radiation

cumulus cloud

normal cloud base

600–1600 m

air bubble

2

sinking air

ascending warm air

4

cumulonimbus cloud

anvil

The clouds which occur most frequently during the summer are the cumulus clouds or the heap clouds. During fair summer days they occur as small white masses in the sky. In favourable conditions they can grow in size and depth into towering thunderclouds.

warm upwinds

cold down-draughts

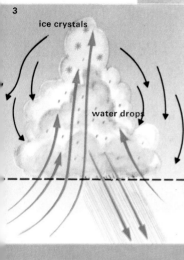

3

ice crystals

water drops

Cumulus clouds are formed by the sun heating the air close to the ground so that it is lighter than the surrounding air. Portions of air move upwards as they become lighter. Pressure drops in the air masses, the air expands and the temperature drops. At a certain level the water vapour condenses and small clouds form (Fig. 1). If the temperature decrease in the surrounding air is fast enough, with height, the cloud might grow in thickness and size. (Figs 2–3) The fully developed Cumulonimbus cloud (Cb-cloud) reaches a height of 8–15 km. The rain is preceded by heavy, cold wind gusts. (Fig. 4)

Cb-Clouds III–IV

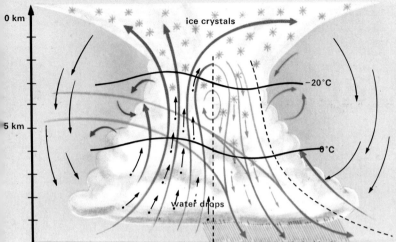

anvil

0 km

ice crystals

−20°C

5 km

0°C

water drops

Strong vertical air currents exist inside the Cumulonimbus cloud. They can amount to 20–30 m/s. In the upper part of the cloud, the anvil, there are ice crystals only while the lower part consists only of water drops. Between the top and the base there is a layer with both supercooled water drops and snow crystals (above). This situation is favourable for the formation of hail since supercooled drops stick to the ice pellets or snow flakes which are lifted up and down in the cloud. The vertical currents are also responsible for the electrification of Cb-clouds, turning them into thunderstorms. A Cb-cloud only lasts for a couple of hours.

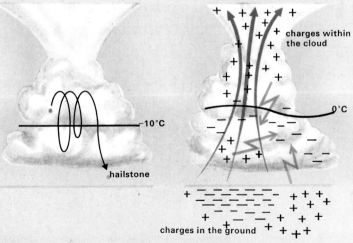

hail

−10°C

hailstone

charge distribution in thunderclouds

charges within the cloud

0°C

charges in the ground

The enormous differences in electric potential within the cloud or between the cloud and the ground is discharged in strong sparks – lightning. The air along the lightning flash is heated explosively thus giving rise to the characteristic thundering noise.

The Ten Main Cloud Types

By observing the clouds in the sky it is often possible to predict the coming weather. Clouds are usually divided into ten main cloud types. Two types of clouds, Nimbostratus and Cumulonimbus, become so thick that they extend right through the whole troposphere. At the bottom of each illustration the latin name of the cloud is given with the popular English name, the height to the cloud base and the symbol used to plot the cloud type on weather maps (see Observations and Weather Maps III–IV).

Cirrus – 'mare's tail'
5–8 km

Altocumulus
2–5 km

Cirrocumulus – 'mackerel cloud'
5–8 km

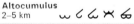

Altostratus
2–5 km

Cirrostratus
5–8 km

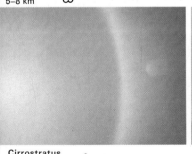

Nimbostratus – rain clouds
2–3 km

Cumulonimbus – shower- or thundercloud
500–1500 m

Stratocumulus
0·5–2 km

Cumulus – 'heap cloud'
500–1500 m

Stratus
0–500 m

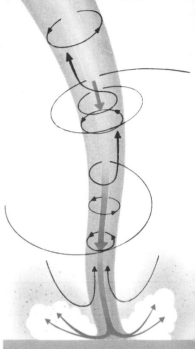

A tornado is a violent whirlwind storm that occurs in many places around the world, but particularly over the densely populated parts of central U.S.A. Tornadoes can be up to 1000 m in diameter and the wind speed may well reach 200 m/s!

𝕽 = thunderstorms

Tornado catastrophe in U.S.A.

Between the 3rd and 4th of April 1974, 80 tornadoes caused the death of 380 people and damage amounting to 2·5 billion dollars.

warm front

rain

L

+22 °C

+2 °C
cold air

Squall line

warm and very
humid air

cold front

10 April, 1947

Tornadoes

A tornado causes almost complete destruction along its path. It moves at a speed of about 20 miles/h (about 32 km/h). The large pressure drop amounting to 50–100 mb (millibars) makes houses virtually explode (right). Trees and cars can be sucked up into the air (above)

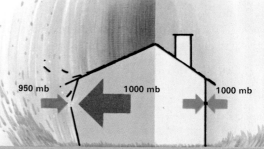

950 mb   1000 mb   1000 mb

Tornadoes form in connection with intense thunderclouds. They often occur in squall lines which form when very warm and humid air is streaming northwards from the Gulf of Mexico ahead of an advancing cold front from the west.

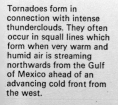

cold air   cold front

warm air

squalls

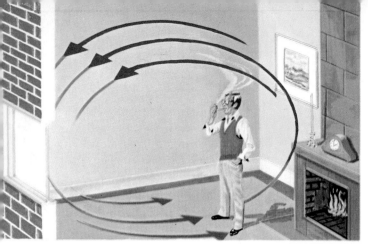

Winds I–II

Local winds arise from air being heated differently in different places. Air that is heated becomes lighter than the surrounding air and rises while cold air is heavier and sinks (above). In this way closed thermal circulations form.

The sea breeze is a wind which blows from the sea towards the land during spring and summer (below). The sun heats the land masses. The hot air rises and cold air from the sea replaces the warm air. During the day the cool sea breeze penetrates further inland. During the night a weaker land breeze forms directed towards the sea (lowest illustration).

In mountain areas, when the large scale weather pattern is fairly settled, local winds may also form. The mountain wind (above) develops when the air along the slopes cools by radiation into space (red arrows); the cold air runs down into the valley bottom (blue arrows).

During the day, however, solar radiation heats the mountain slopes and the air gets lighter and rises. This wind is called valley wind (above). The rising air cools, sometimes producing clouds and, along the mountain ridges, rain showers. A rather special type of mountain wind is the cold, 'katabatic' wind which forms at the edges of big glaciers and over which the air may become very cold and heavy.

Winds III–IV

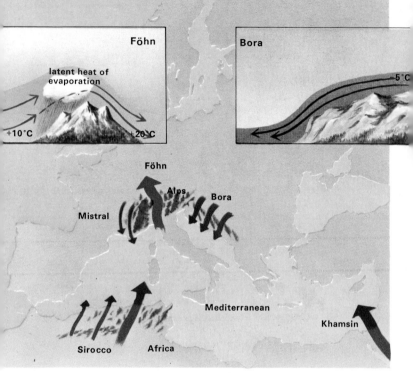

Föhn

latent heat of evaporation

+10°C   +20°C

Bora

−5°C

Föhn

Alps

Mistral   Bora

Mediterranean

Sirocco   Africa   Khamsin

Throughout time people have given names to
local winds and this 'folk lore' is especially
well developed in the Mediterranean area. The
Sirocco, the Khamsin and the Föhn are warm
winds while the Bora and the Mistral are cold
and dry winds. The Bora is an example of a
'katabatic' wind, when cold and heavy air
from mountain areas or glaciers falls down
towards lower lying lowlands or coasts.
Katabatic winds can occasionally reach gale
force. The Föhn winds become dry and warm
from moist air flowing over a mountain range;
in the stable air clouds and rain form on the
windward side releasing latent heat. The heat
produced by this evaporation is added to the
dry air and after passing over the mountains
the air can be 10°C warmer than on the
windward side. Föhn winds occur in many
other places in the world, for example in the
Rocky Mountains and south-east of the
Scottish mountains.

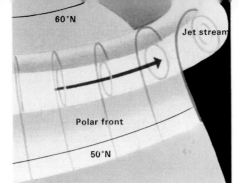

The atmosphere is constantly in motion. At a height of 10 km the wind speed can reach 150 m/s while winds at the surface rarely exceed 45 m/s, i.e. hurricane force winds.

At an altitude of 8–15 km the air is moving very rapidly in so-called jet streams which circle the globe in bands (above and below). The wind speed can amount to almost 200 m/s. The jet streams are coupled to the so-called frontal zones which separate different air masses (above left).

The illustration above shows a photograph taken from the satellite Gemini 12. The band of clouds across the photograph are jet stream cirrus and they show the path of the subtropical jet stream.

Air Masses

Over large areas on earth air can be fairly uniform as regards temperature and moisture. Such air masses and their properties depend on the geographical location and the character of the under-lying surface. One usually distinguishes between three main types of air masses: Tropical air (T), Polar air (P) and Arctic air (A). If an air mass forms over the sea it is called maritime (m) and if over land continental (c). Maritime polar air is denoted by mP etc.

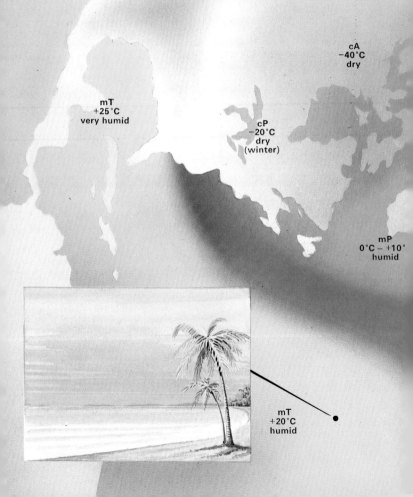

cA
−40°C
dry

mT
+25°C
very humid

cP
−20°C
dry
(winter)

mP
0°C − +10°
humid

mT
+20°C
humid

The main illustration shows different air masses that can affect the weather in Europe. In connection with the migrating depressions different air masses move in over the continent. The depressions usually form in the border zone (the dark blue region in the illustration) between tropical air and polar air. An air mass is also referred to as a warm mass or a cold mass depending on whether the surface of the earth is colder or warmer than the air.

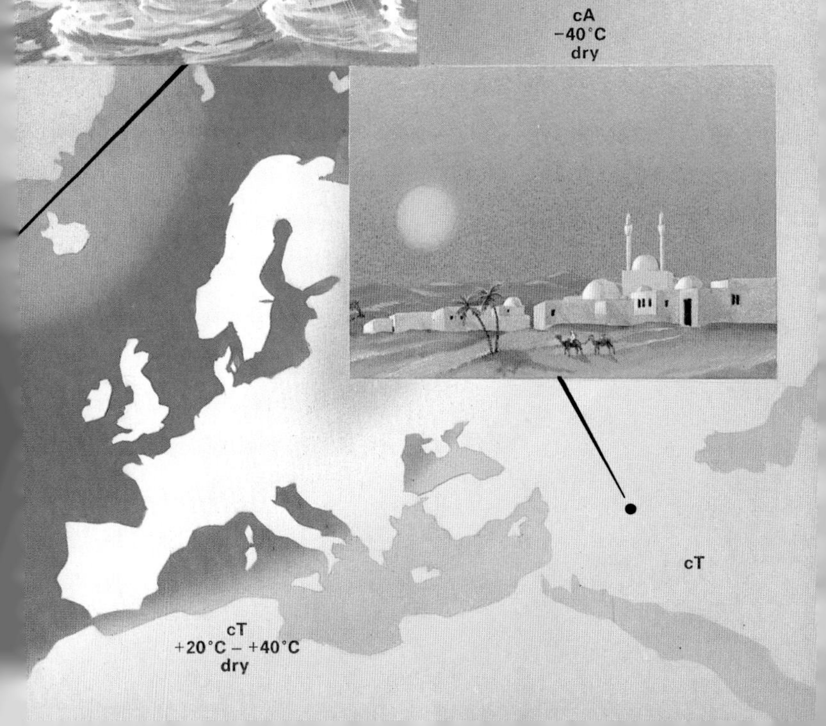

cA
−40°C
dry

cT

cT
+20°C − +40°C
dry

Polar front systems I–II

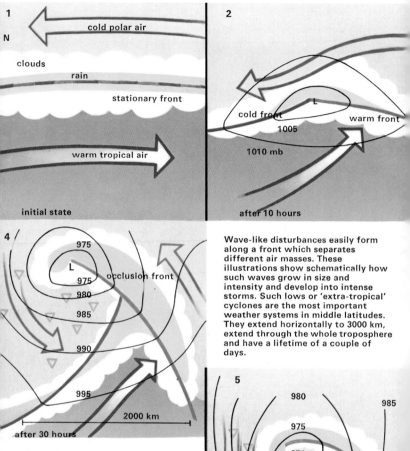

1

cold polar air

N

clouds

rain

stationary front

warm tropical air

initial state

2

cold front

L

warm front

1005

1010 mb

after 10 hours

4

975

L

975

occlusion front

980

985

990

995

2000 km

after 30 hours

Wave-like disturbances easily form along a front which separates different air masses. These illustrations show schematically how such waves grow in size and intensity and develop into intense storms. Such lows or 'extra-tropical' cyclones are the most important weather systems in middle latitudes. They extend horizontally to 3000 km, extend through the whole troposphere and have a lifetime of a couple of days.

In Fig. 4 the polar front cyclone has reached its greatest intensity. The cold front has partly overtaken the warm front, so lifting the warm air aloft, away from the surface. The fronts are said to have 'occluded'. Fig. 5 shows the last stage in the development of the polar front cyclone. The warm air only reaches the ground in the south-east. The central low starts to fill at the same time as a new small wave forms on the trailing cold front. The precipitation along the occluded front has weakened considerably, but numerous rain showers occur in the area of the depression.

5

980

985

975

970

L

C

990

L

995

after 40 hours

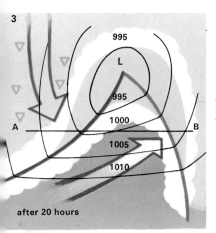

3

995

L

995

1000

A ——————————— B

1005

1010

after 20 hours

▽ = rain shower

In the Figures one can follow the development of a frontal wave as seen on an idealized map, i.e. seen from above. Full black lines show isobars — lines connecting points with the same air pressure. In Fig. 1 a stationary front separates warm tropical air in the south from cold polar air in the north. An area of cloudiness and precipitation extends along the front. Along such a front waves easily form; the warm air pushes towards the north and the cold air towards the south. The new low moves with the wave along the front towards the east with a speed of 50–100 km/h (Fig. 2)

Pressure continues to fall as the depression deepens. The winds around the low pressure increase and ahead of the warm front there is an extended area of precipitation. The cold air behind the cold front advances towards the south with rain or snow showers (Fig. 3).

A
polar air   tropical air   polar air
B

A cross section along the line A–B in Fig. 3 is shown above and another cross section along the line C–D in Fig. 5 is shown below, illustrating the characteristic distribution of clouds in the two cases. In the occluded case below, the cold front has climbed the warm front and the cloud systems have merged.

tropical air

C
polar air   polar air
D

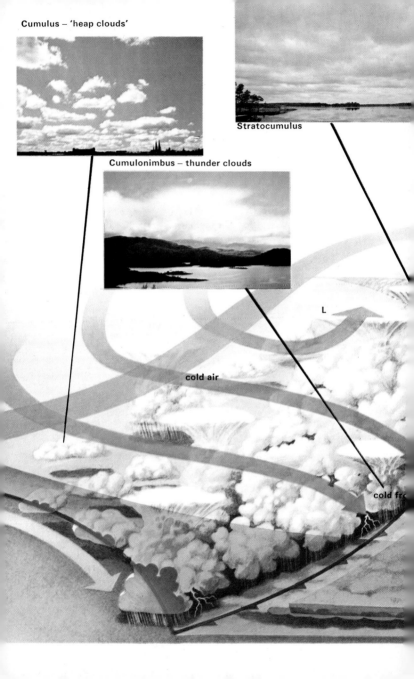

Cumulus – 'heap clouds'

Stratocumulus

Cumulonimbus – thunder clouds

cold air

L

cold fr

Polar front systems III–IV

Nimbostratus – rain clouds

Bird's eye view of a polar front cyclone.

This plate shows an enlarged illustration of Fig. 4 on the previous plate. The warm moist air moves up along the warm frontal surface, cooling at the same time. An extended area of cloudiness and precipitation forms. In the warm sector between the fronts low cloud, fog or drizzle occur. The cold front on the other hand pushes the warm air violently upwards, giving rise to the formation of cumulonimbus clouds. The cold front passes quickly and is followed by clearer skies and windy weather.

Cirrostratus

jet stream

warm front

warm air

cold air

Polar front systems V–VI

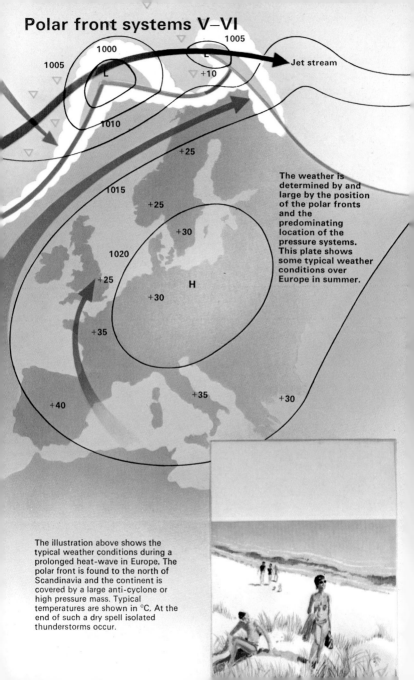

1000
1005
1005
L
+10
L
Jet stream
1010
+25
1015
+25
+30
1020
H
+25
+30
+35
+40
+35
+30

The weather is determined by and large by the position of the polar fronts and the predominating location of the pressure systems. This plate shows some typical weather conditions over Europe in summer.

The illustration above shows the typical weather conditions during a prolonged heat-wave in Europe. The polar front is found to the north of Scandinavia and the continent is covered by a large anti-cyclone or high pressure mass. Typical temperatures are shown in °C. At the end of such a dry spell isolated thunderstorms occur.

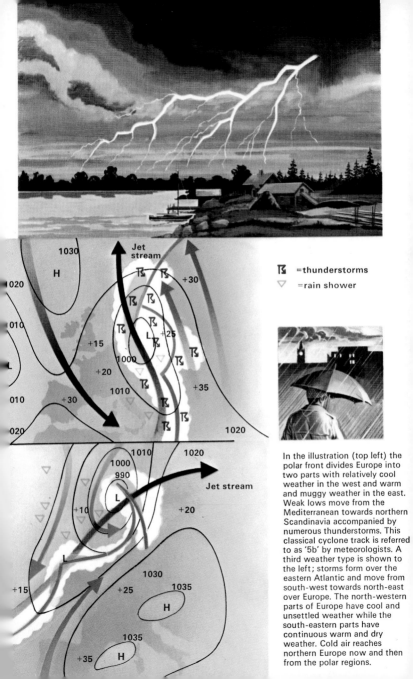

= thunderstorms

▽ = rain shower

In the illustration (top left) the polar front divides Europe into two parts with relatively cool weather in the west and warm and muggy weather in the east. Weak lows move from the Mediterranean towards northern Scandinavia accompanied by numerous thunderstorms. This classical cyclone track is referred to as '5b' by meteorologists. A third weather type is shown to the left; storms form over the eastern Atlantic and move from south-west towards north-east over Europe. The north-western parts of Europe have cool and unsettled weather while the south-eastern parts have continuous warm and dry weather. Cold air reaches northern Europe now and then from the polar regions.

Polar front systems VII–VIII

During the winter the weather is determined to an even greater extent than during the summer by the tracks of the polar front cyclones. During the winter the differences between day and night become fairly narrow in central and northern Europe.

1000

1020
−40
H

1020
−45
H

−10°C

L

0

−5

−5

1000

990
L

+5

Jet stream

+15

1010

The illustration above shows typical mid-winter conditions. Lows move along the polar front over Spain towards Greece carrying cloudy and rainy weather. Northern Europe is covered by very cold arctic air masses.

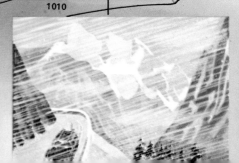

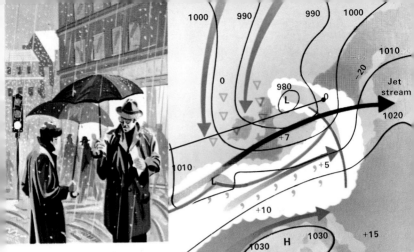

The position of the polar front above brings mild, but unsettled weather to north-western Europe. Lows move from the south-west towards the north-east and behind them mild Atlantic air is brought in over north-western Europe. France and Germany get cloudy weather with occasional drizzle. In the illustration below cyclones are moving from the Norwegian Sea towards the south-west of the U.S.S.R. Cold air is streaming down over north Europe while humid Atlantic air gives rise to widespread areas of fog in the western areas.

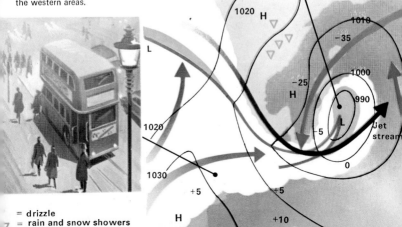

= drizzle
= rain and snow showers

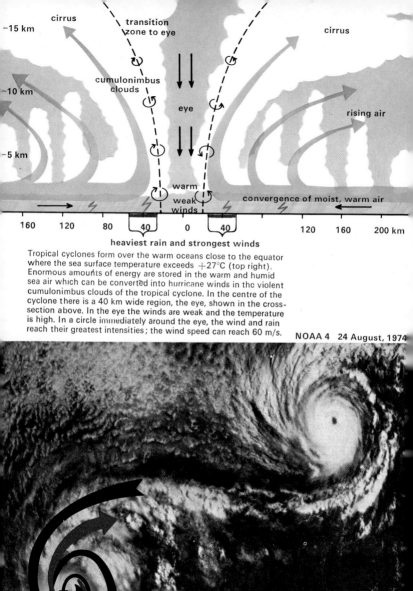

Tropical cyclones form over the warm oceans close to the equator where the sea surface temperature exceeds +27°C (top right). Enormous amounts of energy are stored in the warm and humid sea air which can be converted into hurricane winds in the violent cumulonimbus clouds of the tropical cyclone. In the centre of the cyclone there is a 40 km wide region, the eye, shown in the cross-section above. In the eye the winds are weak and the temperature is high. In a circle immediately around the eye, the wind and rain reach their greatest intensities; the wind speed can reach 60 m/s.

NOAA 4   24 August, 1974

Tropical Cyclones

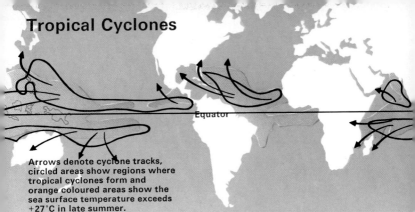

Arrows denote cyclone tracks, circled areas show regions where tropical cyclones form and orange coloured areas show the sea surface temperature exceeds +27°C in late summer.

Tropical Cyclones are called Hurricanes in the U.S.A. and typhoons in Japan and in the Pacific. They are intense storms which hit the coastal areas of countries in these regions with violence. Heavy rain in combination with hurricane force winds and floods causes great material loss as well as loss of human life each year.

An important aid in trying to track down cyclones is the satellite picture. Below left is a unique picture of two tropical cyclones over the Pacific Ocean. Black and red arrows show the air spiralling inwards and outwards respectively at low and high levels.

A man fights against the whipping winds of a tropical cyclone in Palm Beach, Florida. The wind speed is 45 m/s.

The basis for weather forecasts is formed by the thousands of weather observations carried out all over the world which are exchanged through an international telecommunications network.

A weather balloon with radio sonde is released at Sterling, Virginia, U.S.A. at 12 GMT.

Under the weather balloon is a package containing the measuring instruments and a small radio transmitter which continuously send data down to earth about the pressure, temperature and moisture of the air. By following the balloon with a tracking radar it is possible to calculate the wind at different levels from previously recorded positions.

Washington D.C.

A synoptic observation station.

Bracknell, England

radio connection

At a synoptic observation station the thermometers are kept in a special cage. The humidity is often measured by means of a so-called psychrometer.

weather satellite

Sundsvall Airport

receiving station

Swedish Weather Service, Norrkoping, Sweden

Helsink

Observations and Weather Maps I–II

At about 7,000 'synoptic' observation stations a weather observer goes out every three hours to observe the weather. The speed and direction of the wind, the temperature, the humidity and the air pressure is measured and the amounts and types of clouds are estimated. In Sweden alone there are about 200 synoptic observation stations.

telecommunications network

'Main trunk'

Offenbach, Germany

Melbourne

Rome

Special observations are made by aircraft and ships.

All these observations as well as observations from 600 radiosonde stations and observations from aircraft and ships are collected regionally. They are then translated into special numerical codes which can be sent through on the international meteorological telecommunications network. In each country the central weather service is responsible for this. In the U.K. the Meteorological Office in Bracknell, Berkshire, takes care of the collection and exchange of weather observations by the use of communication computers.

Observations and Weather Maps III–IV

Thousands of observations arrive via teleprinters every three hours at weather service departments in the form of strings of numbers. The numerical code is translated into special weather symbols which are plotted on weather maps. The meaning of these different numbers after they have been translated to weather symbols is shown on the right. This particular compact symbol describes the weather in Stuttgart, West Germany, 11 March 1971 at 1200 GMT (circled in the weather map to the right) and is one of several thousands that are plotted on weather maps.

Telex punch tape

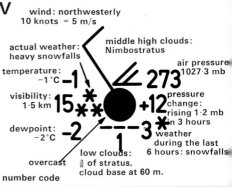

Weather Symbols

wind: northwesterly
10 knots = 5 m/s

actual weather: heavy snowfalls

middle high clouds: Nimbostratus

temperature: −1°C

air pressure 1027.3 mb

visibility: 1·5 km

273

pressure change: rising 1·2 mb in 3 hours

dewpoint: −2°C

weather during the last 6 hours: snowfalls

overcast

low clouds:

$\frac{3}{8}$ of stratus, cloud base at 60 m.

number code

10738 83310 15737 27351 3712/ 5221

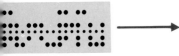

From the plotted weather observations on the weather map the meteorologist marks the different weather systems. The map (top right) is an enlarged section of the map below. The meteorologist draws isobars with the aid of the air pressure observations and depicts the position and limits of fronts and areas of precipitation etc. The complete weather map analysis is shown below.

Analysed weather map
11 March, 1971, 12 GMT.

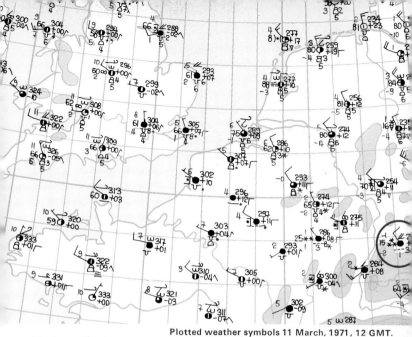

Plotted weather symbols 11 March, 1971, 12 GMT.

Weather satellites have become an important aid in weather analysis especially over the oceans. To the left in the satellite photograph below, the cloud system associated with the fronts over the Atlantic is quite clearly visible.

Photograph from weather satellite ESSA 8, 11 March 1971, 11 GMT.

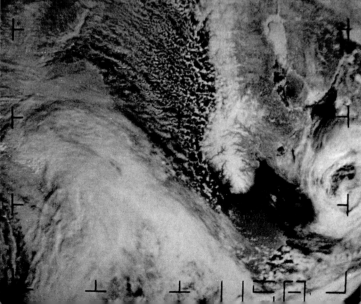

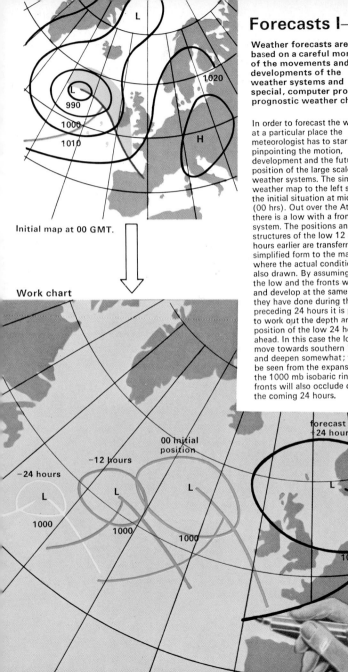

Forecasts I–II

Weather forecasts are based on a careful monitoring of the movements and developments of the weather systems and special, computer produced, prognostic weather charts.

In order to forecast the weather at a particular place the meteorologist has to start by pinpointing the motion, development and the future position of the large scale weather systems. The simplified weather map to the left shows the initial situation at midnight (00 hrs). Out over the Atlantic there is a low with a frontal system. The positions and structures of the low 12 and 24 hours earlier are transferred in a simplified form to the map below where the actual conditions are also drawn. By assuming that the low and the fronts will move and develop at the same rate as they have done during the preceding 24 hours it is possible to work out the depth and position of the low 24 hours ahead. In this case the low will move towards southern Norway and deepen somewhat; this can be seen from the expansion of the 1000 mb isobaric ring. The fronts will also occlude during the coming 24 hours.

Initial map at 00 GMT.

Work chart

Computer forecast

A relatively new aid in weather forecasting is computer made forecast maps. The results of very extensive numerical computations are presented on an electrostatic plotter (above). In this case the computer has predicted the formation of a new frontal wave south-west of Wales, which the meteorologist takes into consideration when constructing his final +24 hour-forecast chart (below).

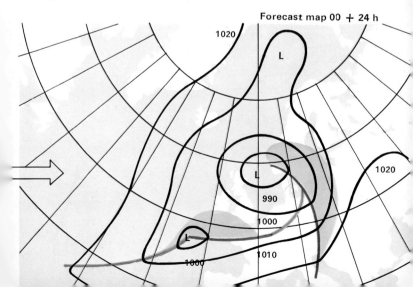

Forecast map 00 + 24 h

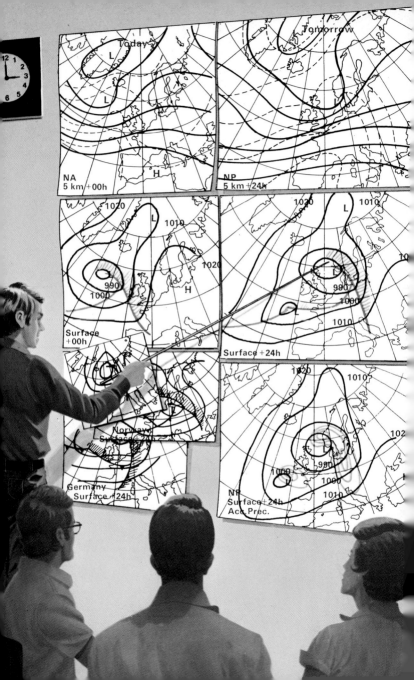

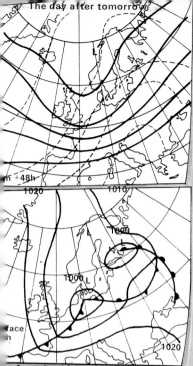

The day after tomorrow

m +48h

1020   1010
1020   1000
1006 L
1020

face
n

A number of weather charts, showing different aspects of the atmosphere and its motion, form the basis of the meteorologist's weather forecasts.

At national weather service stations all over the world meteorologists meet every afternoon to discuss the development of weather conditions during the coming two days (left). The weather forecasters are primarily interested in the flow of the air at the surface and at a height of about 5 km as it is presented on various charts. The broken lines in the upper charts show the so called relative topography which is a measure of the mean temperature in the layer between the ground and 5 km. NA means a numerical analysis and maps are labelled for the initial time (NA +00) produced by the computer. About 4,000 weather observations have been used to achieve this. On the wall there are also computer forecasts (NP) for +24 and +48 hours ahead (NP +24, NP +48), for the surface and 5 km. These maps and the forecast map produced by the meteorologist as well as maps from foreign weather service stations are thoroughly discussed.

Forecasts III–IV

The meteorologist in charge uses these discussions as a basis to adjust his forecast maps and formulates the local weather forecasts for the different districts in the country. Forecasts are disseminated in many different ways to the public. They are broadcast on radio several times a day but weather maps usually make it easier to understand the forecasts presented, for example, in the daily newspapers and on television.

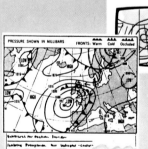

Computer models of the atmosphere have greatly improved forecasts.

The complicated mathematical equation in the middle of this page is actually a very simplified description of the movement of the atmosphere and this equation can be used in conjunction with a computer to make forecasts. The Height, h, to a pressure level changes because of several different processes in the atmosphere. The most important ones are shown in the figures to the right and arrows indicate the corresponding terms in the equation for h. Observations of the height to different pressure levels are carried out every 12 hours at roughly 500 places in the northern hemisphere.

The circulation around a low increases if air streams towards the centre in the same way as the rotation of the ice skater increases when she raises her arms over her head, becoming 'thinner' (= inflow)

$$\frac{\partial}{\partial t}\left(\nabla^2 h\right) + f^2 \frac{\partial}{\partial t}\left(\frac{\partial}{\partial P}\left(\frac{1}{\sigma_0}\frac{\partial h}{\partial P}\right)\right) = \mathrm{J}\left(\frac{g}{f}\nabla^2 h + f, h\right)$$

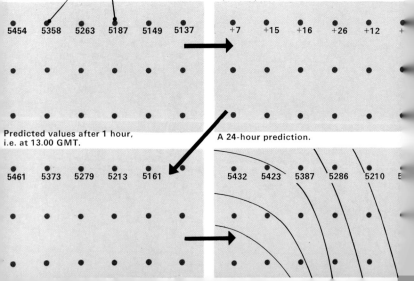

500 mb heights, initial values for 11 March, 1971, at 12 GMT.

| 5454 | 5358 | 5263 | 5187 | 5149 | 5137 |

Height changes in 1 hour computed by the equation above.

| +7 | +15 | +16 | +26 | +12 | + |

Predicted values after 1 hour, i.e. at 13.00 GMT.

| 5461 | 5373 | 5279 | 5213 | 5161 |

A 24-hour prediction.

| 5432 | 5423 | 5387 | 5286 | 5210 | 5 |

Weather Prediction Using Computers

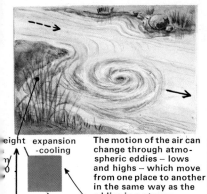

The observed height values are carried over to a regular mesh of calculation points covering the whole northern hemisphere. The number of points amounts to 3000. A section of the mesh is shown lower left. From the measured initial height values for different pressure levels, among others 500 mb (about 5000 m), the computer can calculate changes in the height h in time steps of 1 hour at a time. By repeating these computations 24 times the computer arrives at a 24 hour

Temperatures at a particular place may change through warm and cold air masses being transported in the horizontal layers.

The motion of the air can change through atmospheric eddies — lows and highs — which move from one place to another in the same way as the eddies in a stream.

Vertical movement can create temperature changes.

height
m

expansion -cooling

air bubble

compression -warming

air pressure in mb

0   500   1000

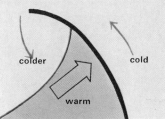

colder   cold

warm

prediction as shown below. Instead of the mesh point values, lines for equal height to the pressure 500 mb — isohypses — have been drawn. The computer has also computed a forecast of the surface pressure and the precipitation for the last 12 hours. About 7 billion additions and multiplications are needed for a 24 hour forecast.

$$g f J\left(\frac{\partial}{\partial p}\left(\frac{1}{\sigma_0}\frac{\partial h}{\partial p}\right), h\right)$$

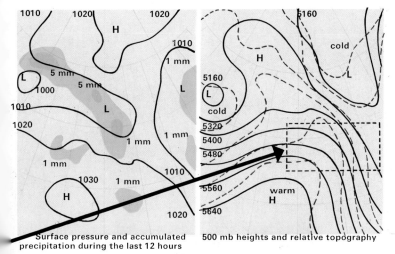

Surface pressure and accumulated precipitation during the last 12 hours

500 mb heights and relative topography

Numerical prediction charts valid for 12 March, 1971, 12 GMT.

Rossby Circulation

cold

cooler and dry

warm and humid

The General Circulation I–II

The atmosphere is like a giant heat engine transporting heat from the tropics to the polar regions.
North of about 40°N the earth loses heat by thermal radiation, while south of the same latitude there is a surplus of radiational energy. In order for the tropics not to become constantly warmer and northerly latitudes not to become colder the excess heat in the tropics must be exported to the cold regions. Between the equator and 30°N this takes place by warm air rising at the equator carrying warm air upward and northward while the air at 30°N sinks and is cooled by strong long wave radiation over the cloudless desert areas, and finally moves towards the equator in the trade winds.

warm

60°N

30°N

Hadley Circulation

Equator

This transport system is called the Hadley circulation. In northern latitudes, where the effect of the rotation of the earth is much greater, this is no longer the most efficient way of transporting heat, however. The motion of the air instead takes place in 7000–10,000 km long so-called Rossby waves. In the northgoing branches warm air is transported towards the north while in the southgoing ones cold air streams southwards. This can be interpreted as a huge stirring and mixing of the atmosphere.

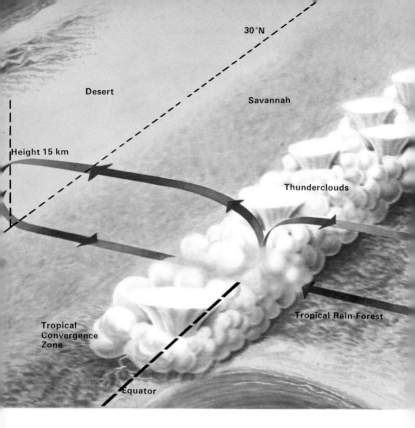

30°N

Desert

Savannah

Height 15 km

Thunderclouds

Tropical
Convergence
Zone

Tropical Rain Forest

Equator

The illustration above shows the northern and southern branches of the Hadley circulation which merge over the equator into one stream of rising air. Along this zone there are numerous, intensive thunderstorms. They can reach up to 18 km height and transport huge amounts of heat and moisture in their strong up-draughts. This zone is found more or less continuously around the whole earth and is called the Tropical Convergence Zone (TCZ); it is the driving force of the Hadley circulation. The warm rising air streams towards the north in the upper troposphere. Over the extended desert areas and in the subtropical highs around 30°N the air is cooled by long wave, infra-red radiation, and sinks. In the lower part of the troposphere there is a return flow in the trade winds of cooler and drier air which, however, picks up heat and moisture on its way back towards the TCZ.

North of 30°N the Rossby waves take over the heat transport but the real Rossby waves have a more irregular appearance than the idealized ones drawn on the previous plate, since they are influenced, for example, by the large mountain ranges. The illustration (top right) shows the flow of air over the northern hemisphere 24 March, 1970. The lines between the different shades of blue show the same height to the pressure 500 mb, approximately at 5000 m. The air moves parallel to these lines, moving faster the closer together they are. The deeper the blue the lower the height. A number of lows in different stages of development move in the general flow of the Rossby waves. At the same time the Rossby waves also move slowly from the west towards the east. The satellite photograph (lower right) shows the cloud and precipitation areas which correspond to the frontal systems in the illustration above it. The cloud band around the equator is made up of the cumulonimbus clouds of the TCZ.

The General Circulation III–IV

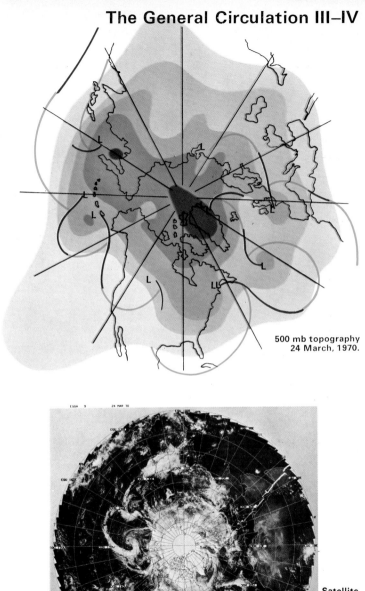

500 mb topography
24 March, 1970.

Satellite
photograph f
24 March, 19

The Hydrological Cycle

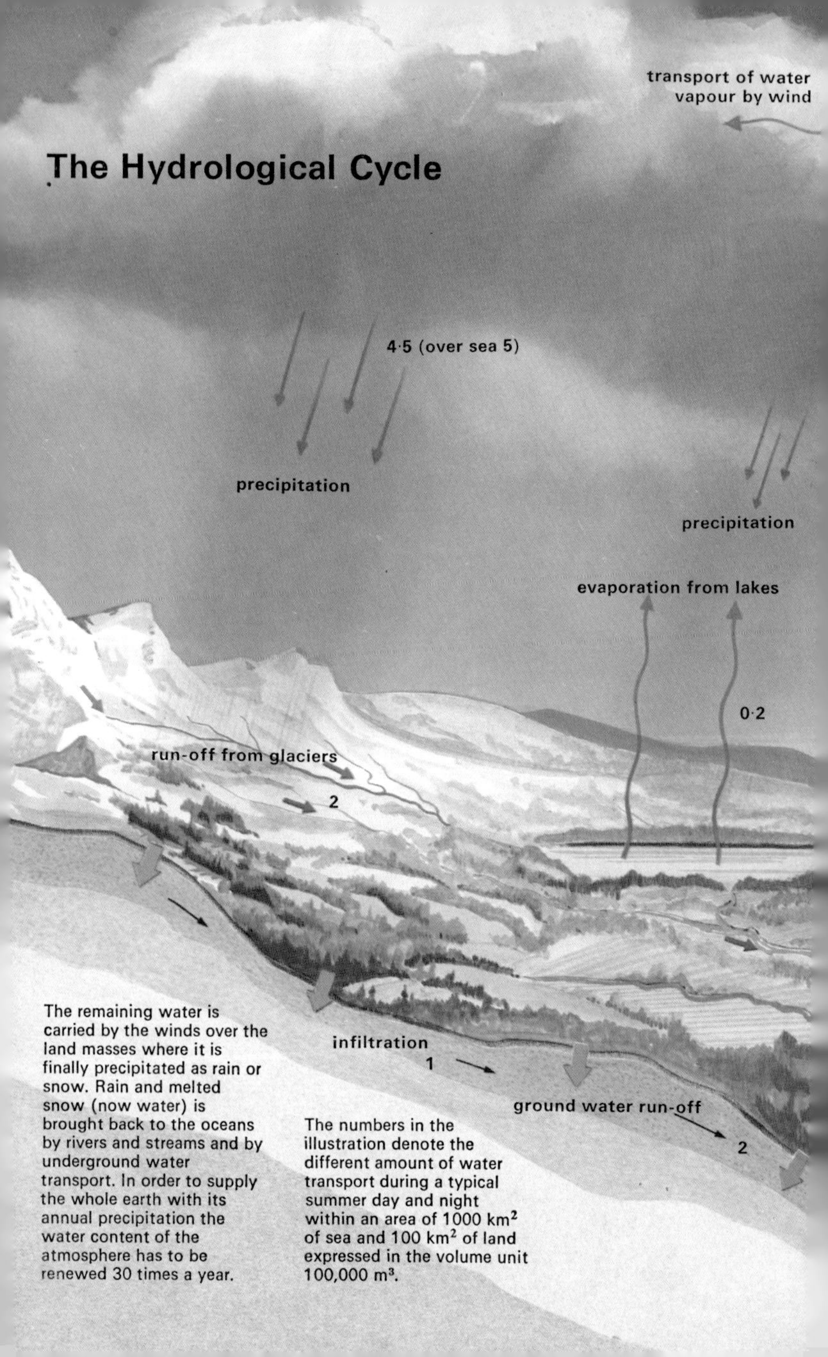

transport of water vapour by wind
4·5 (over sea 5)
precipitation
precipitation
evaporation from lakes
0·2
run-off from glaciers
2
infiltration
1
ground water run-off
2

The remaining water is carried by the winds over the land masses where it is finally precipitated as rain or snow. Rain and melted snow (now water) is brought back to the oceans by rivers and streams and by underground water transport. In order to supply the whole earth with its annual precipitation the water content of the atmosphere has to be renewed 30 times a year.

The numbers in the illustration denote the different amount of water transport during a typical summer day and night within an area of 1000 km^2 of sea and 100 km^2 of land expressed in the volume unit 100,000 m^3.

Water is one of the most important compounds in nature, but it is distributed unevenly over the earth. The heat from the sun is the driving force in the continuous circulation of water called the hydrological cycle.

This heat evaporates large amounts of water from the oceans which are added to the air as water vapour. Evaporation also takes place from land, lakes, rivers and streams as well as from the vegetation — so called evapotranspiration from the stomata of leaves. A large fraction of this water is immediately returned to the seas by rain.

vertical transport of water vapour

evaporation from ground and vegetation   2·4

10

evaporation from the sea

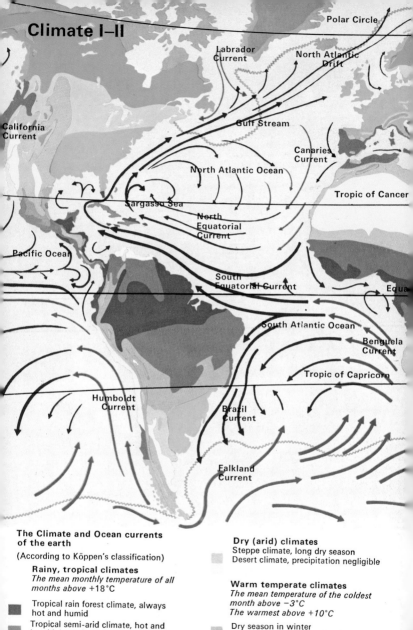

Climate I–II

The Climate and Ocean currents of the earth

(According to Köppen's classification)

Rainy, tropical climates
The mean monthly temperature of all months above +18°C

Tropical rain forest climate, always hot and humid

Tropical semi-arid climate, hot and humid, but with a dry period

Dry (arid) climates
Steppe climate, long dry season
Desert climate, precipitation negligible

Warm temperate climates
The mean temperature of the coldest month above −3°C
The warmest above +10°C

Dry season in winter
Mediterranean climate, dry season in summe
Humid all year round

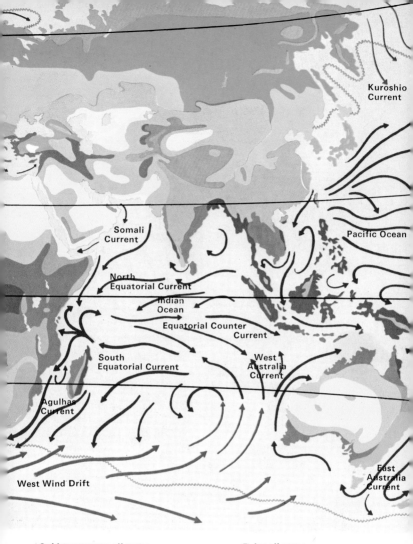

Kuroshio
Current

Somali
Current

Pacific Ocean

North
Equatorial Current

Indian
Ocean

Equatorial Counter
Current

South
Equatorial Current

West
Australia
Current

Agulhas
Current

West Wind Drift

East
Australia
Current

Cold temperate climates
*The mean temperature of the coldest
month below −3°C
The warmest above +10°C*

Humid all year round
Dry season in winter

Polar climates
All months below +10°C

Tundra climate, mean temperature of the
warmest month between 0°C and 10°C.
Glacial climate, mean temperature
of the warmest month below 0°C.
cool or cold ocean current
warm current
The extreme limit of drifting
ice during winter.
(The map above illustrates winter in
the northern hemisphere)

Climate III–IV

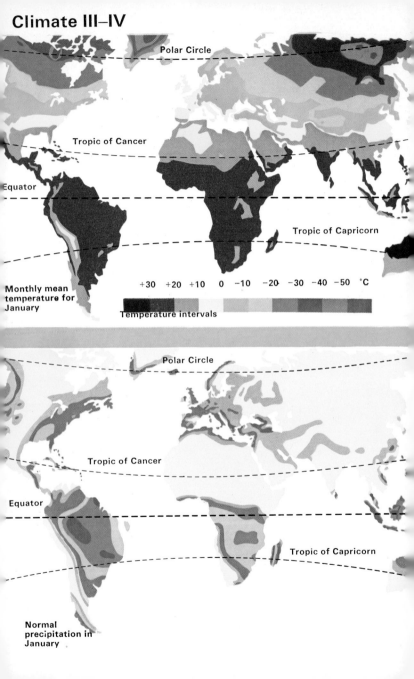

Polar Circle

Tropic of Cancer

Equator

Tropic of Capricorn

+30 +20 +10   0 −10 −20 −30 −40 −50 °C

Monthly mean
temperature for
January

Temperature intervals

Polar Circle

Tropic of Cancer

Equator

Tropic of Capricorn

Normal
precipitation in
January

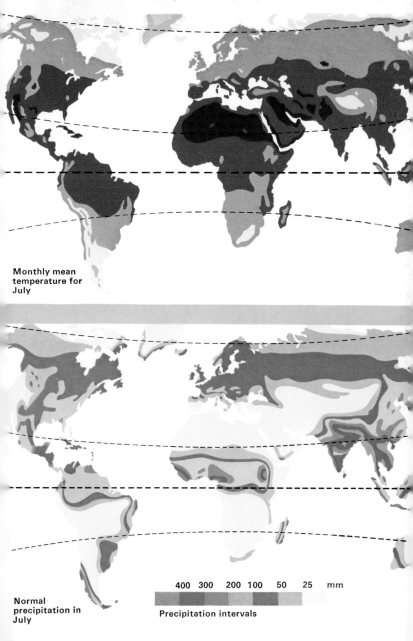

Temperature and precipitation for January and July

Monthly mean
temperature for
July

Normal
precipitation in
July

400   300   200 100   50   25   mm

Precipitation intervals

Climate V–VI

Surface pressure and winds during winter and summer.

The map to the right shows the mean surface pressure and mean winds during January. The key (far right) explains the different symbols. The colour scale shows the intervals of pressure over land. The situation during the winter is dominated by the mighty high pressure over Central Asia and the deep low in the neighbourhood of Iceland. Correspondingly deep lows are also found in the Gulf of Alaska and north-east of Japan.

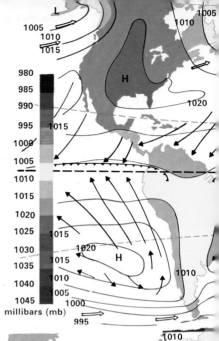

millibars (mb)
980
985
990
995
1000
1005
1010
1015
1020
1025
1030
1035
1040
1045

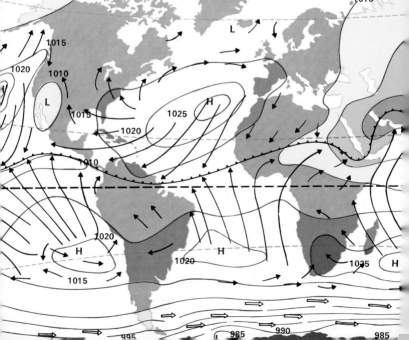

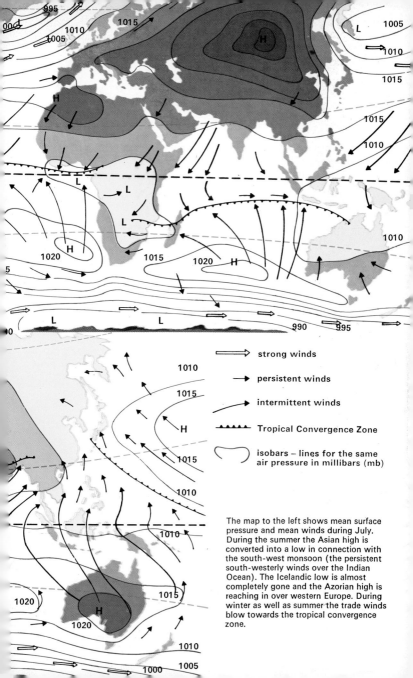

strong winds

persistent winds

intermittent winds

Tropical Convergence Zone

isobars – lines for the same air pressure in millibars (mb)

The map to the left shows mean surface pressure and mean winds during July. During the summer the Asian high is converted into a low in connection with the south-west monsoon (the persistent south-westerly winds over the Indian Ocean). The Icelandic low is almost completely gone and the Azorian high is reaching in over western Europe. During winter as well as summer the trade winds blow towards the tropical convergence zone.

The climate normally varies considerably from place to place, but the activities of man have also started to affect the properties of the air. This is primarily detectable as changes in the local climate. We are increasingly polluting the air by emitting smoke from chimneys and smoke stacks and by car exhausts.

Local climate is to a large extent determined by the geographical location of a particular place in relation to seas, lakes, mountains or hills and valleys. Big cities are different as compared to the surrounding rural areas. In winter or during clear nights with weak winds an inversion, i.e. a layer in which temperature rises with height, often forms by long wave radiational cooling of the ground. This inversion is higher up over the city than over the rural areas and acts as a lid on the vertical mixing of the atmosphere below. Beneath the dome shaped inversion (below) air pollutants accumulate. Heating from houses and the increased mixing over the rougher 'edges' of the city makes the temperature several degrees higher than in the surrounding countryside. In winter the temperature difference can be as large as 10°C.
Also the city is warmer than its surroundings in summer. The strong heating of the air over the city during the day gives rise to a vertical stream which favours the formation of thunder – or shower clouds. In this way an urban area receives more rain than the countryside around it.

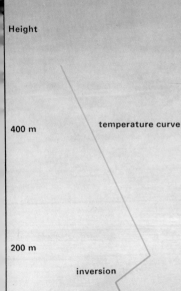

Height

400 m

temperature curve

200 m

inversion

polluted air

0°C   +5°C   +8°C

sulphur dioxide SO$_2$

Carbon monoxide and sulphur dioxide are two of our commonest air pollutants. In the clouds sulphur dioxide is transformed into sulphuric acid which finally gets washed down by the rain. The increased acidity of the lakes has killed much of the vegetation and animal life.

Local Climates and Air pollution

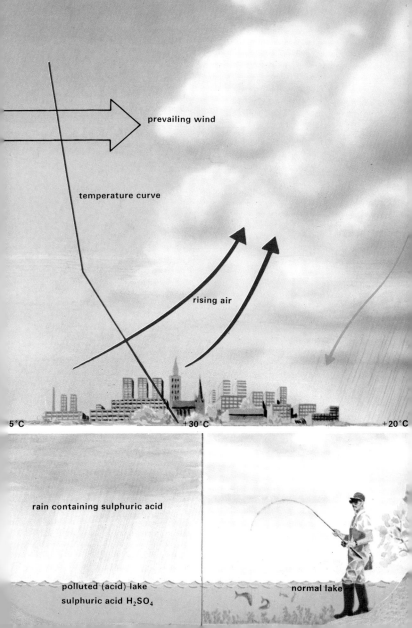

prevailing wind

temperature curve

rising air

5°C   +30°C   +20°C

rain containing sulphuric acid

polluted (acid) lake
sulphuric acid H_2SO_4

normal lake

Project Stormfury

tropical cyclone

silver iodide crystals

Weather Modification

'Everyone talks about the weather, but no-body does anything about it.'
Those words are no longer true even if the possibilities of taming the polar front storms seem minimal now.

At one airport fog is dissolved by means of 15 jet engines. In the Midwest cumulus clouds over fields are seeded with silver iodide, with mixed results.
More serious attempts, as in project Stormfury, are to seed tropical cyclones with crystals of silver iodide in order to stimulate the area of precipitation to grow bigger. Such an enlargement is usually accompanied by a decrease in intensity of the cumulus convection and indirectly in the maximum wind speed. The reduced wind speed also reduces the damage caused by cyclones (larger illustration above).
In the U.S.S.R., where hail very often destroy the crops, anti-aircraft guns are used to shoot silver iodide grenades into hail producing clouds to try to make them produce many small ice crystals instead of large hail stones (left).

silver iodide crystals

cumulonimbus cloud with hail

7 WEATHER OBSERVATION, MAPS AND FORECASTS

In discussing weather forecasts one is entering into the field of practical meteorology. Meteorology as applied to weather forecasting today is based on so-called *synoptic meteorology*. This has its roots in the worldwide network of weather observation stations which regularly provide information about the present weather to meteorologists.

Interest in weather and forecasting goes back a long time in history. The Ancient Greeks recorded weather signs and observations, collating them into a system of rules so that they could predict future weather conditions. Of course they were not always right, but quite a lot of what they were doing was based on sound experience. During the Middle Ages a multitude of experiences and rules about weather and farming, as well as natural events, was drawn up, and in 1508 the first edition of the German 'Farmers Almanac' was printed, quickly becoming very popular in Europe. Sailors too, through years of hard experience, became expert weather forecasters.

A great step forward towards meteorology as we know it today was taken when mercury barometers were first put into use during the seventeenth century. At the beginning of the eighteenth century three thermometers were invented which are still in use today: Reaumur's, Fahrenheit's and Celsius' thermometers. In Berlin temperature records go back to 1719 without interruption and by the middle of the nineteenth century a fairly extensive network of weather observation stations existed in Europe.

The modern weather information service and the exchange of observations was only possible, however, after the invention of electric telegraphy. The first weather service was established in France after an alarming event took place in the Black Sea during the Crimean War against Russia when on November 14, 1854, both the French and English fleets were badly hit by a severe storm. It was later discovered that the same storm had crossed several European countries a couple of days earlier, and the military authorities realized that it might have been possible to predict this storm happening in the Black Sea. In France a weather service was organized, founded on the telegraphic exchange of weather observations which were relayed from the harbours along the coast. These observations were plotted on maps and so the first synoptic weather maps were produced.

Many countries in Europe set up national meteorological services around the middle of the last century; the Meteorological Office in Britain was established in 1854 as a department of the Board of Trade, mainly in response to the need to try and prevent the loss of lives at sea. Synoptic charts, showing the weather situation over as wide an area as possible at a given time, were drawn, using observations sent by telephone or telegram. Such charts are still the basis of modern meteorology, although the collection of data has been greatly amplified and accelerated by the use of radio, teleprinters and satellites.

It was soon realized that international co-operation was essential, and from 1873 onwards, the Directors of the several national meteorological services formed an association which came to be called the International Meteorological Organization (I.M.O.), in order to co-ordinate their efforts.

However, the real progress in meteorology came during World War II. Both Germany and the Allies invested a lot of money in obtaining more accurate weather forecasts. The biggest customer was the Air Force with an increasing demand for weather information to support its operations. International co-operation and the exchange of weather observations suffered of course during these years, but after the war countries soon worked together again, with strengthened resources. The number of stations where regular observations were carried out increased, and in particular those which observed the free atmosphere. Forecasts were also becoming more reliable; civil aviation expanded enormously in the late 1940s and early 1950s and the traffic across the Atlantic Ocean required a good network of observation stations. A number of weather ships were anchored in the middle of the Atlantic to serve civil aviation.

In 1951, the I.M.O. was replaced by one of the United Nations technical organizations, known as the World Meteorological Organization, with its permanent headquarters at Geneva, Switzerland. The member states of W.M.O., acting through their Executive Committee, their Technical Commissions and their Regional Associations, are now responsible for all international co-operation in meteorology at the governmental level.

The 1950s and 1960s have seen two important technical innovations take their place in meteorology. In 1950 the first numerical calculation on an electronic computer of the atmospheric flow at a height of 5 km was carried out. The first computer in the world, Eniac, devoted twenty-four hours of its time to compute a twenty-four-hour forecast over the U.S.A.! It computed at the same time rate as the weather itself, but could not give forecasts. Since then, however, progress has been rapid and the speed of electronic computers has increased ten-thousandfold. The corresponding computer forecast today would take less than a second on a modern fast computer. This development in the capability of computers has been used to produce more complicated and improved mathematical models (theoretical representations) of the atmosphere. The results of these numerical computations are nowadays a natural and necessary component of the work in a modern weather service.

On October 4, 1957, the first satellite, Sputnik 1, was launched by the U.S.S.R.; only the size of a football, it struck the world with amazement. By 1960 the first weather satellite, the American Tiros-I, that could photograph the atmosphere of the earth, had been launched. Satellites have become an important aid to the work of meteorologists. It has become easier to detect and observe weather systems in such areas as over the oceans and in the tropics. We now have satellites orbiting around the poles as well as geo-stationary satellites which move over the equator at the same speed as the earth's rotation, following the same spot all the time. From satellites we can get photographs of the earth's atmosphere both during the day and, by means of the infra-red radiation from the earth and the atmosphere, during the night. We are on the verge of a real breakthrough in getting observations of temperature and humidity from the free atmosphere. This development might result in considerably improved forecasts during the 1980s.

Synoptic observations

So-called synoptic observations are taken simultaneously all around the world every third hour. When making observations all countries refer to the same time, GMT, i.e. Greenwich Mean Time. GMT is also denoted by a Z after the time so that the observations are made at 00Z, 03Z, 06Z, etc. (sometimes written as 0600Z, etc.).

At a synoptic observation station the observer takes measurements of atmospheric pressure, temperature, humidity, wind, both direction and speed, and the change of pressure during the last three hours. The present weather and clouds are also observed and the weather conditions for the last three hours are noted. The clouds are described in terms of low, middle and high clouds and the observer registers cloud amounts and types. Some stations have equipment which measures the height to the base of low clouds, and this is also true for airports which make synoptic observations. The meteorological visibility is estimated by using well-known landmarks in the area.

A synoptic observation station is shown in the colour plate 'Observations and Weather Maps I–II'. On the earth there are about 7000 synoptic observation stations evenly distributed from north to south. In the early days of meteorology people such as teachers, farmers, light-house keepers and railwaymen were recruited as observers. Now many observations are made at airports within the framework of civil aviation. A number of private citizens, often housewives, are employed as auxiliary observers; this means that they have to get up every third hour, day and night!

A synoptic observation station

How is a weather observation done? About twenty minutes before the official observation time the observer walks out to the thermometer screen. This is a ventilated wooden cage on four legs placed in an open, representative place away from houses, hills, trees, etc. to ensure that the thermometers measure the temperature of the air and are uninfluenced by local surroundings. The thermometers are situated 2 m above a grass surface ($1\frac{1}{2}$ m in Britain). There are usually four thermometers in the screen: two ordinary (one 'dry-bulb' and one 'wet-bulb') thermometers, plus one minimum and one maximum thermometer (set at twelve-hour intervals).

Humidity can be measured continuously by a hair hygrometer. Human (preferably blonde) hair extends when the humidity of the air increases and contracts when it is getting drier, so that the length of a small bundle of hairs may be used as a measure of the humidity. A more accurate instrument for the measurement of atmospheric humidity is known as a *psychrometer*, consisting of the two ordinary thermometers, one a 'dry-bulb' thermometer and the other a 'wet-bulb' thermometer. The latter has a dampened muslin cloth wrapped around its mercury bulb, and from the resulting drop in temperature due to evaporation it is possible to compute the dew point and relative humidity of the air using Hygrometric Tables. Atmospheric humidity is normally reported in terms of *dew point*.

Minimum and maximum temperature

At the times of o6Z and 18Z the observer reads the maximum and minimum temperatures, using two especially constructed thermometers. The maximum thermometer works like an ordinary clinical thermometer, with a narrow passage close to the bulb which prevents the mercury flowing back into the bulb once the highest temperature during the period has been reached. The minimum thermometer on the other hand is an alcohol thermometer, since mercury freezes at a temperature of $-39 \cdot 87°C$. In the thin glass tube containing the alcohol there is a small metal bar shaped like a small dumb-bell. The minimum thermometer is kept horizontal so that when the alcohol column retracts the small bar moves with it because of the surface tension at the free end of the column. If the temperature rises, however, and the alcohol column moves up the scale the bar remains at its lowest position. The lowest temperature during the twelve-hour period can be read from the position of the bar; after being read both thermometers are re-set.

Clouds and weather

While the observer is out at the temperature screen he is also recording the actual weather. Rain, snow, thunder, hail, mist, drizzle and tornadoes are some of the ninety-nine different weather types he has to report. These are subdivided into different intensities and the weather phenomena are described as being continuous or intermittent, using an internationally agreed numerical or alphabetical code.

The clouds form an important part of the standard weather observation. During the day it is comparatively easy to judge the presence of different clouds. During the night on the other hand it is not an easy task to try to distinguish between twenty-seven different cloud types in three different layers. Finally the observer also estimates the visibility of known distances to landmarks in the area around the station.

Air pressure

Back indoors the observer measures the air pressure, generally by means of a mercury barometer. At some places the simpler aneroid barometer ('aneroid' meaning 'non-liquid') is used. The aneroid barometer (Fig. 27) is made up of a metallic box that has been partially emptied of air. The corrugated lid of the box is very

Fig. 27   An aneroid barometer. The small, adjustable needle makes it possible to read changes in pressure. Many synoptic observation stations, however, are equipped with the more accurate mercury barometer. Most aeroplane altimeters are aneroid barometers and height is marked in metres or feet, instead of mb by means of the relation between height and pressure shown in Fig. 2.

thin and moves in and out when the air pressure rises or falls. A needle shows these movements on a dial measured in mb. The aneroid barometer is the simplest and most common barometer, but is not as accurate as the mercury barometer used in meteorology. The mercury barometer consists of a vertical glass tube, sealed at the top, filled partially with mercury and erected over a small mercury reservoir. The air pressure over the reservoir makes the top of the mercury column go up and down and air pressure is read from the height of the mercury column. The older unit for pressure, inches of mercury, derives from these barometers (inch Hg). Nowadays we use millibars instead. Observation stations are at different altitudes, so that in order for the meteorologist to be able to compare different pressure observations the air pressure, as measured by a mercury or an aneroid barometer, has to be reduced to a common level – that of sea level. If you have a barometer and want to compare your readings with those of the weather stations or with the weather maps produced by the weather service you should adjust your barometer to show the pressure at sea level. The required information is obtainable from your nearest weather station.

Wind

Some observation stations are not equipped with wind gauges and the wind has to be estimated by the observer by, for example, noting the effect of the wind on smoke, trees, flags, etc., according to special instructions. Most official reporting stations are equipped with wind gauging devices, the most common one being the cup anemometer which is mounted on a mast 10 m above the ground (see Fig. 28). The wind that is reported from a weather station is usually the average wind during ten minutes, i.e. the mean wind speed. At airports more accurate and precise wind information is needed by aircraft at landing and take-off. The ordinary anemometer is comparatively slow in its response to wind variations, i.e. gusts, and airports are therefore equipped with special gust wind meters capable of measuring these fast wind variations. Wind speed is reported in knots.

Fig. 28 Wind gauges, anemometers, are often of the 'cup' type. A wind vane measures the wind direction. Normally the anemometer is placed 10 m above ground level.

Pressure change

Another important piece of information is how pressure has changed – the pressure tendency – during the last three hours. This is measured by means of a *barograph* which is an aneroid barometer where the bulging of the box is transferred to a pen which traces the pressure on a revolving paper chart wrapped around a drum. From this graph, showing the variation in time of pressure, the observer can read off the character and amount of pressure change.

Precipitation

At times o6Z and 18Z the observer empties the rain gauge, a standard metal bucket standing away from houses and obstacles. The bucket collects the rain and snow falling within a twelve-hour period. The amount of precipitation is measured in a measuring glass and is reported in mm (millimetres) or inches.

At some stations measurements of snow cover are carried out and at coastal stations sea swell is measured. At big climate and synoptic stations measurements of sunshine duration and radiation are also made.

At airports all around the world observations are made every hour, and at the big international airports they are even made

every half hour. These weather observations are especially designed to serve the great need for accurate weather information at
landing and take off.

Ship observations

Over the vast oceans it is difficult to man observation stations, but
it is still very important for the meteorologist to get information
from these areas. Since shipping companies also have a vested interest in weather observations and weather forecasts over the sea,
forty countries have joined in equipping more than five thousand
ships with the facilities to carry out synoptic observations. The
ships make their observations every sixth hour when at sea. Mer-
chant ships, however, follow well-trafficked trade routes which do
not always coincide with the areas of greatest interest to meteorologists, as for example, the North Atlantic. Civil aviation requires
information about the weather from these areas, however, and it
was particularly important in the days before jet aircraft. Through
the international civil aviation organization, I.C.A.O., nine
weather ships permanently anchored at different positions in the
Atlantic Ocean were paid for. These ships make observations
every hour and measurements of the free atmosphere every sixth
hour. Because of economic reasons, however, these nine ships
have been cut down in numbers to only four. One of them is the
weather ship Mike (4YM), also referred to as the ship Polarfront
by the Scandinavians, which is anchored in the Norwegian sea.
This ship is of great importance in detecting weather systems
approaching Europe from north-west.

From a prognostic point of view the number of observation
stations in the Atlantic is very unsatisfactory and many erroneous
forecasts occur because of incomplete knowledge about the
weather systems over the Atlantic Ocean.

Radio soundings

The construction of the radio sonde supplied meteorologists with
a fairly simple and cheap tool for carrying out measurements of
the state of the atmosphere above the ground up to high altitudes.
The radio sonde is a balloon, inflated with hydrogen, carrying a

small instrument package. The rubber balloon is capable of lifting this package up to an altitude of 25,000 m (or the equivalent of roughly 75,000 feet). In the instrument package are a simple aneroid barometer, a hair hygrometer and a so-called bimetallic thermometer. The bimetallic thermometer is based on the principle that two different materials expand differently when heated. The thermometer consists of two thin metallic strips of different materials in the shape of a coil; when the temperature changes the coil expands or retracts. The information from these three instruments is continuously sent down to the earth by means of radio signals.

The height of the balloon can be computed from the simultaneous temperature and pressure measurements.

At most radio sonde stations the balloon is also followed or tracked by radar. The radar gives information about the height of the balloon and the slant distance to it. From these data the drift of the balloon and the average wind can be computed for different layers.

Radio sonde observations are made twice a day at about 700 locations in the northern hemisphere. Balloons are released at noon and midnight G.M.T. every day. The release of a weather balloon from the radio sonde station in Sterling, Virginia, U.S.A. is shown to the left in the colour plate 'Observations and Weather Maps I–II'.

The network of radio sonde stations is much less dense than the network of surface stations. As we have seen, the air flow is a very important factor in weather forecasting and the problems with observations from the ocean areas are even greater for radio sonde observations. In order to fill these gaps over the oceans meteorologists have great hopes for the development of satellites capable of measuring pressure, temperature and moisture content in the free atmosphere by remote sensing. Such measurements are already available, but are not as accurate as radio sonde observations.

Pressure or height?

In order to see the motion of the air close to the surface we study the distribution of pressure by drawing isobars from the (sea level) pressure observations plotted on a map. In the free atmosphere above the ground, however, it is easier and more convenient to measure the height to a given pressure rather than observing the pressure at a given height and then drawing isobars for that height. The pressure 500 mb is found at roughly 5 km height, but it is higher in high pressure areas than in low pressure areas. If we draw maps with lines connecting points with the same height to a specific pressure, say for example 500 mb, these lines would have the same properties as isobars on a surface map. The air would move parallel to these lines and the wind would be stronger where the lines are closer together. The rules for the air motion around a high and a low would also be the same as for isobars (see Chapter 4). Lines connecting points with the same height to a specific pressure are called *contours*. The contours are basically the same as the topography lines on a geographical map showing the height of the terrain. Sometimes meteorologists actually refer to the 'topography of 500 mb'.

The pressure surfaces 500 mb and 300 mb have special importance in the work of weather services. Since normal pressure at the ground is about 1000 mb we have equal amounts of air above and below the pressure 500 mb, which then acts as a kind of average of the whole atmosphere. The 300 mb chart is particularly important to modern civil aviation since it shows the jet streams at the cruising levels of modern jet airliners. The 500 and 300 mb charts are prepared in most places by computers and the information disseminated to the pilots via the meteorological services at the airports.

Since World War II other sources of weather information have been brought into use. Aircraft flying across the Atlantic report about the weather en route, wind speeds and temperatures. They are often equipped with weather radar and can thus give accurate information about the weather systems through which they pass.

Radar

Radar (*Ra*dio *D*etection *a*nd *R*anging) is used extensively in meteorology. By using pulses of electromagnetic radiation with a wave length of 3–8 cm it is possible to get echoes from large cloud and rain drops. This gives the possibility of detecting and following the movement of areas of precipitation on the radar screen. Cumulonimbus clouds contain large drops and become visible on the radar screen. Since the cumulonimbus cloud has a short life-history and develops rapidly, radar is a very valuable complement to the synoptic observations in predicting local weather. By tilting the radar antennae in the vertical it is also possible to estimate the height of the tops of cumulonimbus clouds and areas of precipitation. Radar has been installed principally at big airports in order to supply information to the pilots and the aviation meteorologists.

Weather Satellites

The weather satellite is perhaps the most spectacular piece of modern technology meteorology has taken into use during recent years. A large number of satellites have already been launched – and died – and substantial improvements in the instruments of weather satellites have taken place since the first one, TIROS-I, was launched in 1960. Both the U.S.A. and the Soviet Union have sent up satellites and Europeans are working at present on putting a weather satellite into orbit over the Atlantic at the equator in 1978. This project is called Meteosat and will provide pictures both of the weather over most parts of the Atlantic and Europe as well as radiation measurements.

A weather satellite is shown in the colour plate 'Observations and Weather Maps I–II'. The American satellites are orbiting around the poles at an altitude of about 1400 km, passing the same spot on the earth twice a day. They take pictures of the earth and the atmosphere in both visible light and infra-red making it also possible to observe the clouds during the night. The pictures are bounced down to earth by means of a television camera. The system is automatic and anyone on the earth's surface who is equipped with a radio receiver can pick up the radio signals and transform them into photographic pictures; this system is called APT (Automatic Picture Transmission). At the present time NOAA-4 and 5 are in space supplying us continuously with pictures in both visible light and infra-red. Infra-red is the same as the heat radiation from the air and the earth's surface. Since these pictures react to different temperatures it gets easier to determine the types of clouds observed on the pictures. An example of a satellite picture is shown in the colour plate 'Observations and Weather Maps III–IV'.

The Russian satellites are of the METEOR series. They move at a considerably lower altitude, about 600 km, and the pictures cover smaller areas, but give better resolution to the cloud masses. The Soviet satellites have not been equipped with APT and the pictures have not been made available to other countries.

The satellites in the NOAA-series as well as the new satellites which will be launched over the years to come are equipped with an instrument that makes it possible to measure accurately the heat radiation from different layers of the atmosphere. By knowing the amount of radiation in different wave-length intervals it is possible to calculate the vertical distribution of temperature in the atmosphere. This is what arouses the hopes of meteorologists for the future. These satellite observations already give some additional information over the oceans, but the satellite observations are not of the same accuracy and quality as the radio sonde observations. They cannot therefore yet replace the latter, but the chances are very good that satellite technology will develop up to the point where meteorology will be able to supply us with a wealth of new observations all over the globe.

Besides these regularly observing satellites there are many others which have special applications. The ATS and SMS satellites are geostationary satellites hanging at fixed positions over the equator and have partly meteorological applications.

Exchange of weather observations

All the weather observations made around the world would be of very limited value unless they could be exchanged rapidly and communicated between the weather services of the world. A synoptic weather observation is written in an internationally agreed code, i.e. written as a row of numbers, that can be sent in a compact form as a telegram, by radio or on telex. In the colour plate 'Observations and Weather Maps I–II' the means by which this exchange of weather observations is carried out is illustrated.

The complex system for exchanging weather observations and other data between different countries requires well-oiled international co-operation. In order to solve problems of this character an international organization has been established – the World Meteorological Organization, W.M.O. – in which nearly every nation is a member country. W.M.O. is a technical agency of the United Nations, similar to the World Health Organization, W.H.O.

There is a continuous supply of a very large amount of data that has to be taken care of, and in order to deal with them as quickly and efficiently as possible computers have to be used by the weather services in most countries.

From observations to weather maps

The thousands of weather observations reaching the weather service every third hour are the foundation of its forecast work, from reports of actual weather to weather forecasts up to five days ahead. When the observations reach the weather service they are in the form of endless lines of numbers. How they are treated after that is explained in colour plate 'Observations and Weather Maps III–IV'. The numbers are 'translated' into weather symbols which are plotted on maps; these symbols give the whole weather observation in a compact form and are easy to read after some training.

For example, the cloud amount in the weather symbol is depicted by different filling of the station circle. The different cloud types are represented by the cloud symbols given in the colour plate 'The Ten Main Cloud Types'.

The small line sticking out from the weather symbol ending in one or more barbs denotes the wind speed and its direction. One long barb means 10 knots (kts) while a short one means 5 knots. When the wind speed is more than 50 knots it becomes inconvenient to add all the barbs and instead a filled triangle symbol has been introduced for 50 knots.

Air pressure, visibility, temperature and dew point are given directly in figures. This is also true for the pressure tendency, plus a symbol for its character.

Both the present and past weather is given by means of simple weather symbols. Figure 29 is a summary of the symbols used.

As in the observations the weather on the weather map is subdivided into classes of different intensities. Three stars, for example, denotes heavy snowfall while two stars stands for light continuous snowfall. One star means light, intermittent, snowfall. The same rules are valid for the rain dots. A bracket to the right of a weather symbol denotes that the weather concerned has occurred within the last hour, but not at the actual time of observation. In the lower right corner of the weather plot is a symbol for the weather between the time of the former and the present observation – the 'past weather'.

Weather maps

An ordinary weather map comprises something like 2000 weather plots or observations. It is then the task of the meteorologist to bring order to all this weather information by mapping the weather systems hidden in the scattered observations. His work with the weather map is called 'weather analysis'.

The meteorologist draws isobars from the pressure observations. Each observation is a sample of the state of the atmosphere. Between two observations on the map we do not really know what the weather is like, but in general the weather systems are much bigger than the distance between two observing stations. Because

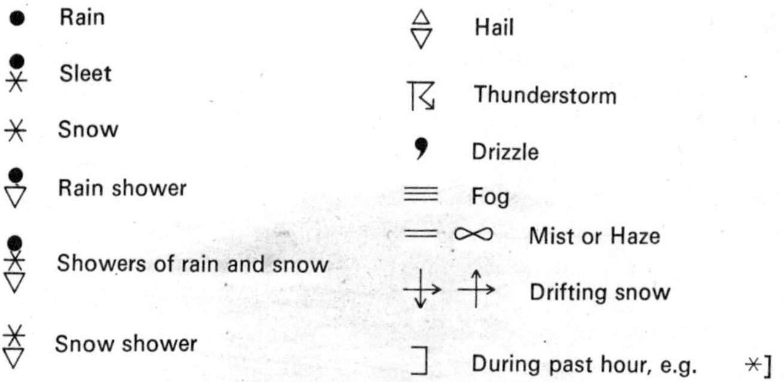

Fig. 29 Weather symbols used on a weather map.

of this there is a fair chance to assume that it is raining between two observation stations if both stations report rain. The average distance between the observations on the map is of the order 50 km over land, but may be 500 km over the sea. Areas of precipitation are easy to map on the chart from the observations and are usually shaded green on the map. It is also easy to distinguish between cloudy and clear skies from the station circles. By drawing lines connecting points with equal pressure – isobars – the meteorologist can understand the large-scale circulation of the air. Together with all the other weather information on the map and from previous knowledge of the structure and behaviour of atmospheric systems the meteorologist can finally draw fronts on the map and separate different air masses from each other. In the lower left corner of the colour plate 'Observations and Weather Maps III–IV' an analysed weather map is shown from March 11, 1972, 1300 GMT.

Out over the oceans satellite pictures are of great help in pinpointing the location and extent of weather systems. The satellite picture in the lower right corner of the Plate shows clearly the frontal clouds north-west of the British Isles where the number of observing stations is small.

The weather map is the meteorologist's most important tool, and a new map is produced every three hours. The final chart is also called an analysis (of the actual weather) in contrast to prognostic charts of future weather patterns that the meteorologists also make.

At a normal weather service station a great number of charts are produced which show different aspects of the state of the atmosphere. The surface analysis shown in the colour plate 'Observations and Weather Maps III–IV' is an overall view taken over a fairly large area. Such a map is suitable for making forecasts up to twenty-four hours ahead. A more detailed presentation of the weather is needed for local forecasts, and in most countries detailed maps showing the weather over a smaller area are prepared. In this way it becomes possible to study and follow the local influences on the weather as affected by, for example, mountains, lakes, big cities, etc.

Charts for movement in the free atmosphere are prepared automatically by computers. The data obtained from radio sonde observations go directly into the computer of the weather service where they are processed by special programmes. The end result is a complete chart with contours or isobars containing other information also of interest to the meteorologists, such as isotherms or isotachs-lines for equal wind speed. These charts are drawn automatically by mechanical or so-called electrostatic plotters or presented on a television screen. In the colour plate 'Forecasts I–II' the upper right illustration shows how a computer chart is drawn by an electrostatic plotter. The weather service produces a number of other charts of interest to various people. Precipitation maps show the rainfall and snowfall distribution over the country for the periods o6Z–18Z and 18Z–o6Z. Every morning and night the minimum and maximum temperatures are presented on special maps. During the winter many people are interested in the thicknesses of the snow cover. In northern Europe the weather services also serve the ships by giving information about the ice situation in coastal regions and at sea by issuing ice charts. These are sent to the ships and icebreakers by means of facsimile.

Forecasts

The 'art' of forecasting would be impossible if the atmosphere did not show a certain regularity in its behaviour. Fortunately, however, the motion of the air follows well-known physical laws often expressed in a mathematical form. Meteorologists base their forecasts on these physical laws valid for the atmosphere, on observations of how weather systems move and develop, and on a careful monitoring of their movements in a series of weather maps, both high up and at the surface. Since the beginning of the 1960s computer forecasts have given valuable guidance to meteorologists.

When making a forecast the meteorologist starts by carefully studying how the weather systems have moved and developed over a major portion of the northern hemisphere for the last few days. He or she tries to get a picture of how the jet streams at a height of 10,000 m are behaving and what changes have and will take place. This gives valuable information about where the cyclone tracks are to be found. Are they in a southerly or in a northerly position? Are cyclones likely to affect us or not? Is the general weather pattern stationary or can great changes be expected during the forecast time? Is there a chance of intense polar front cyclones forming over the Atlantic? After having tried to answer these questions with the help of all the weather charts available at the weather service station the meteorologist turns to studying the weather system on the actual weather map in somewhat more detail. Colour plate 'Forecasts I–II' shows how the meteorologist works out a forecast chart twenty-four hours ahead. The simplest method is called *extrapolation* which means that the motion of a weather system is extended on the basis of its previous motion. To construct the final forecast map computer forecasts are also used as well as the meteorologist's own accumulated experience of the motion of atmospheric systems.

The forecast map prepared in this way describes the expected large-scale weather situation. But the general public wants weather forecasts for specific locations or areas. In order to come up with the final forecast the meteorologist has to 'translate' the large-scale weather forecast into local weather, by paying attention to local factors that may substantially alter the large-scale weather pattern. For instance, the presence of the Atlantic Ocean, commonly referred to as the 'Gulf Stream', has a pronounced warming effect on the winter weather of the western coasts of the British Isles, so that snow and frost are far less frequent than elsewhere; some areas such as the tips of Cornwall, the Scilly Isles and southern Ireland are almost frost-free. A range of higher hills, such as the Welsh mountains and the Pennine range can effectively alter the weather from the windward to the lee side. On a smaller topographical scale, minor variations in the height contours can result in large differences regarding the incidence of fog and frost.

When the meteorologist formulates his forecast he must consider all such influences and base them on his experience of local conditions.

At many places all over the world statistical methods are being used in attempts to assess the local weather from forecast maps containing information about the future shape of isobars and frontal positions. Records of millions of weather observations going back up to sixty years are stored on magnetic tapes and can be used to study the normal behaviour of local weather in different weather situations. The computer can handle a vast amount of historical material in the form of numbers, and extracts necessary information which is stored in its 'memory' in the form of 'tables'. Based on the pressure forecast over a large area prepared by the meteorologist or by the computer itself the computer can come up with local weather forecasts for different districts and presents tables of forecasted temperatures, winds, clouds and precipitation. These computer forecasts have become a valuable aid to the meteorologists, and it is even possible to prepare completely automatic computer forecasts where the computer directly prints out the worded forecast district by district in plain language!

Neither the computer forecast nor the forecast of the meteorologist is on its own of the same quality as that which the combination of man plus machine can produce. At the present time the automatic forecasts are only used internally by the meteorologists, but that might not be true in ten years time if we continue to go on improving and developing our computer models.

Computer or numerical forecasts

What, then, is a computer forecast? The migrating cyclones and anticyclones at our middle latitudes have a lifetime amounting to a few days. It is therefore possible to follow and extrapolate the motion of a low, say twenty-four hours ahead. When we come to longer forecasts the weather is not decided so much by the weather system on the actual weather map, but to a large extent by weather systems *not* existing at the time of the preparation of the forecast. The ordinary methods for working out a twenty-four hour forecast can no longer be used as the basis for the longer forecasts. It is for forecasts up to five days that computers and computer weather prediction have become extremely important.

Computer forecasts are extensive calculations by numbers. By doing that we are utilizing the knowledge of the physical laws for atmospheric motion. These are expressed in the form of mathematical equations that are solvable only in numerical form, i.e. by numbers representing the atmosphere. These very complicated equations require the solving of vast numbers of computations.

For example, something like seven billion simple multiplications, divisions and additions are needed to calculate a twenty-four hour forecast for the northern hemisphere. An attempt at such a calculation was undertaken in 1911 by the British meteorologist L. F. Richardson, long before the appearance of electronic computers. He carried out ten thousands of simple arithmetic operations in order to arrive at the pressure change at the surface for a place in the middle of Germany. It took him five years to complete this work. When he finished, it turned out that his result was completely wrong. Richardson's 'failure' unfortunately cooled down the interest for predictions of this kind for several decades to come. However, in 1950 three scientists were able to prove that computer forecasts were really feasible in practice. In fact Richardson was basically right in his ideas but had run into problems that still have not been completely solved. The first computer forecast in 1950 was carried out for a region of the 500 mb surface over the U.S.A. The computations took twenty hours for a twenty-four hour forecast on the ENIAC computer at Princeton and was a great achievement. What had earlier been a utopia for scientific investigations is now routine at most weather service stations around the world. In very complicated and comprehensive models of the atmosphere scientists are even trying to simulate the climate of the earth and to find an explanation for climatic changes.

Colour plate 'Weather Prediction Using Computers' demonstrates how a computer forecast is made. The computations are nowadays carried out on several levels in the troposphere in order to improve the accuracy of the forecast. The heights to different pressure levels are observed every twelfth hour around the globe and serve as the starting point for the computer. The observations are, however, irregularly distributed over the earth, as we have seen previously, and the computer has to start by transferring these data into a regular network, a lattice with equal distances between the mesh points. Such a regular net is also called a grid and it takes about 3000 grid points to cover the whole northern hemisphere if the grid points are 300 km apart. A low with a diameter of 3000 km will then be represented by about 100 grid points.

The computer presents the results of its calculations in different maps. The lower right corner of the colour plate 'Weather Prediction Using Computers' shows the resulting twenty-four hour forecast for the pressure 500 mb. The computer has also calculated the mean temperature between the ground and 500 mb. From the figure one can see that the low is cold and the high is warm. The second map shows isobars at the surface twenty-four hours ahead and the accumulated precipitation during the last twelve hours of the forecast.

For civil aviation purposes the computer prints out forecast charts showing the jet streams and the wind speeds at 9000 m or 30,000 feet. The airline companies use these jet stream charts in the planning of their flights. The upper winds are especially valuable for long-distance flights. By using the wind forecasts the pilots can avoid, for example, strong head winds when flying to North America and take advantage of the strong tail winds when flying back, and an Atlantic flight might in this way take up to two hours' less time. The passengers arrive more quickly at their destination and the airline companies save costly fuel.

Five-day forecasts

Most weather services prepare outlooks for the coming five days. When working out a five-day forecast it is necessary to take into account the interaction between the polar front cyclones and the long Rossby waves. The meteorologist faces the problem of predicting lows and highs that do not exist at the time of issue and it is not possible to formulate the five-day forecast as precisely as the shorter twenty-four hour forecast. Middle-range and long-range forecasts were only feasible after the advent of electronic computers and the development of numerical models of the atmosphere. Since the computer carries out its calculations for the whole northern hemisphere there is also the possibility that it may catch some of the sophisticated interactions which lead to displacements of the jet stream and changes of its strength and consequent changes in the cyclone tracks.

In Britain, the further outlooks given at the end of the general weather forecasts usually only refer to the subsequent two or three days, but reviews of the weather prospects, giving the general character of the expected weather, though not the precise timing, covering the next seven days, are put out on the special television programmes for farmers.

It is customary for national services to exchange their computer-based forecast weather maps by a radio or teleprinter system known as *facsimile*. Ships, equipped with the necessary receivers, can also consult these charts.

Computer forecasts must be looked upon as approximate suggestions regarding the future development of the weather. It is not uncommon for different computers to arrive at different results for the same forecast period and the meteorologist must rely on his own experience to decide in which one to believe – if any. In many places long lists of different weather 'cases' have been stored. It seems natural to try to go back in time to find an earlier case in the history of weather which resembles the actual weather situation, so that it might be possible to get some hints about future weather developments by checking how the atmosphere developed last time. This method, put into system, is called the *analogue method* and has been used mainly for forecasts extending beyond five days. The atmosphere very rarely repeats itself exactly, however, and one has to be very careful when applying the analogue method.

Most numerical forecasts beyond the first two days only show the flow at 500 mb and when the meteorologist has come to a decision about the general development of the large-scale flow he must in some way translate this into forecast maps for the surface pressure showing the positions of lows and highs and frontal systems. Also when making five-day forecasts the computer uses all the information about earlier weather which has been stored on magnetic tapes. On the basis of how the weather used to be at different flows at 500 mb the computer prints out tables with suggested temperatures, precipitation and winds for a number of places in the country. All the different computer forecasts are 'interpreted' in this way. The meteorologist chooses and discards, adjusts and changes, and finally formulates the five-day outlooks. In the final analysis it is the experience and knowledge of the meteorologist which decides the formulation of the forecast. The five-day outlook contains a general description of the expected weather situation with an indication of cyclone and anticyclone tracks. Forecast temperatures and precipitation amounts are given as mean for the whole period with a hint of expected changes over the five-day period, for example, that the beginning of the period will be warmer than the end.

It is not possible at the present time to give more details than this, but the increased capability of computers will probably give rise to the production of improved models of the atmosphere, resulting in more detailed medium-range forecasts. The most important development will, however, be forecasts for up to ten days ahead.

The map discussion

Every day, in practically all weather service stations around the world, the forecasters meet to discuss the weather developments. The duty meteorologist who has been working with the forecast maps all day presents and makes an introductory discussion of all the maps, satellite pictures and films that are available, especially the ones he has been making direct use of, such as those shown in the colour plate 'Prognosis III–IV'.

The meteorologist on duty discusses his forecast for the different districts, the local forecasts, which will be sent to radio and television stations. A discussion follows leading to minor corrections or changes. Long discussions can occur when the longer range forecasts are discussed.

Long-range forecasts

Predictions for longer times than two weeks are usually called long-range forecasts. In U.S.A., England, Germany and the U.S.S.R. monthly forecasts are made regularly and in some places even forecasts for a three month's season are prepared.

It is impossible to describe the weather day-by-day for such a long period ahead. Instead the long-range forecast tries to give an idea of the average temperature and the accumulated precipitation expressed in fairly crude classification: warmer than normal, normal and below normal, and correspondingly for precipitation.

Even forecasts of this kind can be of great value to many clients such as electric power companies, who can plan the total output of power for the whole coming month. Farmers can plan their work and society as a whole can use long-range forecasts in other kinds of planning.

The long-range Weather Prediction Unit of the National Meteorological Service in Washington D.C., U.S.A., issue twice-a-month forecasts for the coming thirty days.

An example of a British long-range forecast is shown in Fig. 30.

Predictability

Why is it so difficult to make longer forecasts? As we have seen the interactions between different scales of motion in the atmosphere are very complicated. When forecasts are extended beyond a week it is no longer enough to consider only what happens in the northern hemisphere. Weather events in the southern hemisphere can also cause reactions in the northern hemisphere. The motion of the atmosphere also interacts with the oceans. The sea currents and the ocean surface temperatures have been shown to have a marked influence on weather and the positions of the jet streams over extended periods.

Another important factor has to do with the fact that the atmosphere in some places and at some times can be in a state of delicate balance and small impulses can knock it off this balance. With such a situation, it is very difficult to judge in what direction the atmosphere will move. In the case of a polar front cyclone, a

SECTION I—WEATHER PROSPECTS IN THE UNITED KINGDOM FOR THE COMING MONTH
DECEMBER 1977
(Issued on 30 November 1977)

The next 30 days are expected to be marked by contrasting weather types, with periods of mild, fairly wet weather and mainly southerly winds, but also some very cold northerly outbreaks.

Over the month as a whole this is likely to result in mean temperatures and total rainfall near average in all areas.

Gales, fog, frost and falling snow will probably all occur with about average frequency.

Fig. 30 An example of a long-range forecast (reproduced by permission of the Controller of Her Majesty's Stationery Office).

small harmless wave on a polar front might, within a day or two, develop into an extensive and intense storm capable of drastically changing the weather pattern over a large area.

Are there other disturbances which have the same effects? Yes, there are. A small cumulus cloud might under suitable circumstances grow through the whole troposphere into a massive thundercloud. A thundercloud usually lasts only for a few hours, but it can still have a decisive effect on future weather conditions over a much bigger portion of the atmosphere than suggested by its own size if the atmosphere is at a 'turning point'. It is sometimes impossible to forecast a thundercloud an hour in advance, much less two weeks in advance. The effect of each 'unpredictable' disturbance like this is spread out like the waves from a stone dropped in the middle of a pond. After a couple of weeks the effect might have spread over the whole northern hemisphere and the atmospheric flow might look very differently from what it had been if the thundercloud or the polar front cyclone had never occurred. This is what limits the length of time for our forecasts. The Professor of Meteorology at Massachusetts Institute of Technology, Edward Lorenz, states the same ideas even more dramatically by saying that even the flaps of the wings of a seagull might irreversibly alter the flow of the atmosphere in due time; the seagulls are something we cannot control. Another American scientist expresses it a little differently: 'It is not possible to make a forecast for a weather phenomena beyond the normal lifetime of the phenomena in question.' Monthly forecasts would then only be possible if there are changes or rhythms in the weather with periods of about a month forcing the atmosphere back to the narrow path despite all the disturbances which try to knock it off balance.

When working with monthly forecasts one needs to look at previous weather events. In England the Meteorological Office has retained weather maps since 1888 to have material available for comparison with current weather situations. It would be an unreasonable task to look through all these maps manually and so use is made of computers. The computer has all the previous weather events stored on magnetic tapes, but even with computers it takes quite a time to examine all the maps for some ninety years and since the forecast time is as long as a month one has to be very strict about what is a 'similar' case.

The meteorologist has to make use of all types of statistics and knowledge about the atmosphere when working on long-range forecasts. The relationships between sea surface temperatures between different places, between pressure at different places, the upper air flow and what is typical at different times of the year, all have to be considered. Computers are still of limited use when dealing with monthly forecasts, and no computer exists today capable of executing the calculations necessary to produce a forecast for a month in the same way as for five days ahead.

European co-operation

The biggest and fastest computers become overwhelmingly expensive for one country to buy. Because of this sixteen European countries have joined together in a big project, aiming at delivering ten-day forecasts by 1980. By means of the fastest computer in the world, a CRAY-I, it will be possible to carry out the enormous amount of calculations required for a ten-day forecast. The computing centre, the full name of which is the European Centre for Medium Range Weather Forecasts (ECMWF) will be located near Reading in Great Britain. In these extended forecasts the centre will try to make day-by-day predictions of the movements of weather systems. Many people and institutions in society are in need of longer forecasts and it is hardly necessary to point out what ten-day forecasts will mean to holiday makers who will be able to plan holiday trips and weekend excursions with more confidence.

Weather forecasts are made available to the public in many ways. In Britain, for example, morning newspapers carry the general and regional forecasts for the day in question, sometimes accompanied by a weather map; evening newspapers carry forecasts for the following day. Every radio channel puts out forecasts, in either abbreviated or more detailed form, at frequent intervals throughout their transmission times. Television programmes, often presented by the forecasters themselves, give visual representations of the expected weather, on a short twelve- to twenty-

Fig. 31   The forecast districts of the United Kingdom (reproduced by permission of the Controller of Her Majesty's Stationery Office).

four hour basis, and also include the forecast isobaric maps which are so valuable to those who understand meteorology. There are six weather centres, specially designed to deal with enquiries from the general public, situated at Glasgow, Newcastle, Manchester, Nottingham, Southampton and London. In addition there are some thirty forecast stations at airfields which can be telephoned to obtain the local forecast. Because there is a limit to the number of enquiries that can be answered personally by telephone, there is also an Automatic Telephone Weather Service, which provides a recorded forecast for some twenty-three limited areas including the main towns and cities and the favoured holiday areas.

How good are forecasts?

That weather forecasts are not perfect is common knowledge, but they are not as bad as most people tend to believe. The important thing with forecasts is that despite the fact that they are not always right they are still of great value to people who know how to use them properly. How should we then use a forecast? The best way to look at a forecast is to regard it each time as the most reasonable and probable judgement of the future with due regard to the information available at the time of issuing the forecast. From that point of view it is wrong to do what many people do: assume that weather is going to be exactly the opposite to what the weathermen say. On the contrary the wisest way of using the forecast is always to assume that the forecast is right – but when it goes wrong be prepared to change one's mind rapidly! Since forecasts are right on average seven or eight times out of ten you will be able to make much better plans by following the advice of forecasts than by guessing at random.

Forecast accuracy can only be expressed as a vague figure and the sad fact is that it is difficult to measure exactly how good they really are. However, meteorologists have tried for a long time to come up with different scores or measures of the skill of the forecasters, to be able to check afterwards how good the forecasts have been, a process known as *verification*.

Verification

A frequently used measure of verification is the percentage of 'hits'. Let us see how good we are at forecasting rain or no rain at a specific place. By counting the number of times it has rained after rain was forecast and the number of times no precipitation occurred after dry weather was forecast and dividing the sum by the total number of cases, we get the percentage of 'hits', i.e. = (Number of correct forecasts)/(total number of forecasts). If forecasts are of any value we should get a figure above 50%, which represents the results from sheer guessing. Perfect prediction is 100%, i.e. all 'hits'. If we consider temperature instead, things become more complicated because temperature can take on any value from $-50°C$ to $+45°C$. It would be unfair in most circumstances to reject a temperature forecast which was out by only half a degree. Instead we can divide the temperature into sections, for instance 0 to 5, 6 to 10, 11 to 15, etc., and check how many forecasts fall into the right section. The same method can be used with wind forecasts.

Verification of five-day forecasts

What would be the result if we made the simplest of all forecasts: weather conditions will be the same tomorrow as today? Since anyone can make such a forecast the meteorologists also compare their forecasts with this rough method. In Sweden, for example, the score for five-day forecasts based on this method is 62% for temperature and 55% for precipitation, which is better than sheer guessing! This is only a reflection of the fact that the atmosphere has a certain 'habit'. Precipitation varies more than temperature and therefore gets lower figures. It is clear to see, however, that the meteorologists easily beat the 'same for tomorrow as today' method of forecasting.

Verification of computer forecasts

The numerical forecasts are also verified. These do not immediately produce temperature and precipitation forecasts, but mostly numbers which represent pressure or winds over a large area. Thus, a different method from the percentage of 'hits' must be used. By calculating how the forecasted surface pressure varies in comparison with the actual, observed pressure, we get a widely used method of comparison called *correlation*. Correlation can also be expressed in percentages where an entirely correct forecast gets 100% and an entirely wrong one gets 0%. If we look at the surface pressure forecast for twenty-four hours ahead, in Sweden again, for the year 1973 the correlation turned out to be 80% for the whole year. For forty-eight hour forecasts the result was 76%, somewhat less than for twenty-four hours.

However, the ordinary weather forecasts for the coming day are more difficult to verify, because they are expressed in words which can be difficult to interpret exactly afterwards – even by meteorologists. The percentage of 'hits' should be somewhere between 80 and 90% according to some tests.

Accuracy of observations at sea

Insufficient observations are available from large ocean areas and even if the observations point to the presence of a bad weather system out on the Atlantic Ocean they might still not give enough information on the intensity and the speed of the low. The situation has, in addition, deteriorated during the last years because of the withdrawal of some weather ships from the Atlantic. Satellite pictures have compensated a little for this but the situation is really grave. The hope is now for rapid improvement of satellites capable of measuring the heat radiation from the surface and the atmosphere giving us information on temperature, humidity and pressure from different levels, though these measurements are so far not of the same accuracy as those from radio sondes.

What can we expect of the future?

Satellite technology is developing rapidly and it is hoped that within three to four years new satellites will give us observations from the oceans to an extent never seen before, which might eventually lead to a substantial improvement in the quality of forecasts. A general rearrangement of the present, conventional net of observations as well as a faster exchange of observations between countries are part of W.M.O.'s programme for the future. These improvements are also part of a big project called W.W.W. – World Weather Watch.

Bigger and faster computers will be developed even if progress does not continue at the speed it was in the 1960s. Every new computer allows the meteorologists to add more to our knowledge of the atmosphere making it possible to improve forecasts and facilitating detailed studies of still obscure weather processes ranging from climate studies to the growth of fair weather cumulus. Meteorological research within G.A.R.P. – The Global Atmospheric Research Programme – is working along these lines.

We expect the forecasts to be not only better, but also longer and more detailed, though weather forecasts will never be perfect since the atmosphere is too complicated. When Isaac Newton formulated his famous laws and principles for the motion of the planets based on gravity he made it possible to carry out calculations for the future positions of the planets around the sun with almost uncanny precision. We can predict a lunar eclipse or a sun eclipse several thousands of years in advance, but such a prognostic calculation only includes nine planets and a limited number of moons. A prediction of the future state of the atmosphere is more like a calculation for millions and millions of planets all interacting with each other in a constantly changing way.

8 GENERAL CIRCULATION, CLIMATE AND CLIMATIC CHANGES

All the weather systems we have studied so far, cumulonimbus clouds, tropical storms, polar front cyclones and Rossby waves are all parts of the complex atmospheric heat engine. The main task that the air motion is carrying out is to transport the excess heat in the tropics to northern latitudes.

The atmosphere always seems to be trying to find the best and most efficient way of reaching a state of equilibrium. In order to understand how this is done one has to look at the atmosphere as a whole and see how the different parts work together. The motion of the atmosphere as a whole is called the *General Circulation*, which determines where the polar front cyclones, the dry subtropical high pressures and the tropical rain areas are to be found. It is also the key to understanding climate and its variations from place to place. The climate is the accumulated, mean description of weather over a long period of time, and it is usually described in terms of averages of meteorological elements over periods of thirty years. The typical temperature conditions at a certain place can be presented as the monthly mean temperatures. Normal annual precipitation total is the average of a thirty years' series of observations. Variations about such averages also form part of a climatic analysis.

In most places weather changes appreciably from day to day. In contrast, the thirty years' climatic averages usually change very slowly. Nevertheless, climate apparently underwent many dramatic changes in pre-historical time, so radically altering the conditions for life on our earth.

The last ice age, which ceased about 10,000 years ago, is clear evidence of such a change. The twentieth century has, from a climatic point of view, been a golden age. Not since the great 'heat wave' 5000–7000 years ago has the climate been as mild and favourable to humanity. It is not surprising then that concern grows when signs appear indicating that climate may be now starting to deteriorate. The extended drought in the South Saharan border regions has been interpreted as the first step in a general decline in climate and other signs seem to point in the same direction. What kind of role might man himself play in this change? We know that during this century we have substantially affected the composition of the air, the oceans and the biosphere which might in the end influence the rhythm of nature. We are, for example, gradually changing the composition of the atmosphere by releasing a number of compounds which naturally exist only in small amounts. At the present time carbon dioxide, nitrous gases, freons and sulphur dioxide are the main additives. In the same way we add, over limited areas, large amounts of heat. Man has started to manipulate the weather in order to extract water from clouds in regions where little rain is falling and is spreading silver iodide over tropical cyclones to try to diminish the effect of the strongest, hurricane-force winds. The cumulative effect of all these encroachments on the atmosphere on a large scale is practically unknown.

During recent years considerable attention has been drawn towards these activities and the question has been raised whether man is digging his own climatological grave. In order to be able to judge the effects of human activities on the atmosphere at large and locally, we must first understand the causes for the natural changes of the climate which we know have occurred periodically. A prerequisite for that is, of course, that we recognize the different factors working in the general circulation that ultimately shape the climate of our planet. After that it would be possible to ask ourselves which factors are the most important in climatic changes and how would it be possible to control them?

The general circulation

The motion at large over the earth is not so different from that in a pan of water where we heat the outer walls and cool the water in the centre. Actually a number of experiments have been carried out using such a simple device in order to study the fundamental ways of atmospheric motion. In reality the earth is a rotating sphere, but by putting the pan on a rotating disc it is possible to imitate fairly realistically the conditions on earth.

The ultimate source for the motion in the atmosphere is solar radiation. As we have seen in colour plate 'The Atmosphere VII–VIII' there is a net gain of heat south of 38°N and north of 38°S while north and south of these latitudes there is a net loss of radiation. The deficit is largest at the poles and the surplus greatest at the equator. In order for the equator not continuously to become warmer and the poles colder the surplus heat south of 38°N must be moved northward.

The Hadley circulation

The sun is heating the air at the surface in the equatorial regions, while the air over the poles is cooled by outgoing long-wave radiation. The heated air over the equator becomes lighter and starts to rise; the cold air in the north starts sinking which causes the rising warm air in the south to move northward to replace it. At the same time the sinking cold air must stream towards the south to replace the rising air over the equator. We have already seen an example of circulation of this kind, driven by a temperature contrast, but on a smaller scale. This was the mechanism leading to sea and land breezes. In the world circulation warm air flows towards the north in the upper atmosphere, cooling at the same time, while cold air moves towards the south, being heated during its travel. In this way the atmosphere tries to eliminate the temperature differences between the poles and the equator. A circulation of this kind was first suggested by the British scientist George Hadley in 1735, and observations in the tropics have given a certain support to Hadley's ideas. Air really does rise at the equator and moves towards the north in the upper troposphere, but the northward movement ceases at about 30°N where the air begins to sink. A return flow of colder air takes place in the lowest layers of the troposphere in the *trade winds*. The *Hadley circulation* thus only extends to about 30°N (and 30°S in the southern hemisphere) (see colour plate 'The General Circulation I–II'). North of 30°N the atmospheric motion is instead characterized by predominantly westerly winds. We have in fact a westerly wind belt extending all around the globe.

The Tropical Convergence Zone

The rising air stream over the equator is not uniform in character and the transport of heat and moisture upwards from the earth's surface is mainly concentrated in a relatively narrow zone around the earth referred to as the *Tropical Convergence Zone* (*TCZ*). It is illustrated on the left page of colour plate 'The General Circulation III–IV'. Along TCZ there are numerous cumulonimbus clouds and thunderstorms. The cumulonimbus clouds extend up to heights of 15–18 km and transport large amounts of heat and moisture upwards in their strong up-draughts. The cumulonimbus clouds often join together in smaller or bigger clusters of cumulonimbus clouds and thunderstorms. Between these clusters there might be fairly clear and dry weather. The tropical convergence zone is not fixed in the same position, but undergoes continuous changes and displacements towards the north and south. Wavelike bulges towards the north can form and move from the east towards west. The strong cumulonimbus cloud clusters usually form in these waves. The cloud system of the TCZ is visible on the satellite picture in the colour plate 'The General Circulation III–IV' and in the picture of heat radiation from the earth in the colour plate 'The Atmosphere VII–VIII'. Sometimes two TCZs can be observed.

The subtropical jetstream

The TCZ is the driving force in the Hadley circulation. It acts as a big pump that brings hot and humid air up to high altitudes where it moves towards the north and south in the two branches of the Hadley circulation. The earth rotates with a speed of one revolution every twenty-four hours from west towards east; this means that a particle in a fixed position above the surface at the equator moves with a speed of 1670 km/hour. Air rising at the equator and moving towards the north in the upper troposphere will gradually get a velocity surplus as compared with the surface, which increases the farther to the north it comes. Because of this it will move eastward. At about 30°N the air particle would have a speed relative to the surface of the earth corresponding to 230 km/hour or 64 m/sec. It is in these regions we find the subtropical jet stream. It circles the earth in a rather unchanging manner, but at places reaches very high speeds, more than 120 m/sec. The air in the Hadley circulation does not reach further than this. As this air is cooled, it gets heavier and starts sinking. The ceaselessly sinking air gives dry and cloudless weather and over land it creates the dry weather over extensive deserts around the tropics of Cancer and Capricorn such as the Sahara. Even if the surface air is heated very strongly over the desert sand, no clouds form because of the dryness of the air. Therefore it very rarely rains over the desert areas. Out over the oceans the sinking motion takes place in the subtropical high pressure cells, such as the Azores high in the North Atlantic (see colour plate 'Climate V–VI').

The return flow towards the equator of cooler and drier air starts in the subtropical highs. The rotation of the earth makes the air deviate towards the west thus creating the northeasterly trade winds north of the equator or TCZ.

Corresponding trade winds are found in the southern hemisphere. The air flows over a gradually warmer sea, getting heated and so picking up moisture. The trade winds extend up to about 2 km. Above that height air slowly sinks and the cloudiness in the trade winds is in general small, but increases towards the equator and the TCZ. When the air reaches the TCZ it is again hot and humid and ready once more to whirl up violently in one of the thousands of cumulonimbus clouds constituting the tropical convergence zone.

The northern and southern hemispheres differ in one important aspect. The percentage of land in the northern hemisphere is bigger than in the southern, which is almost entirely covered by water. The land masses are heated more easily than the oceans because land has a lower heat capacity than water. On the other hand more heat can be stored in the oceans. They react more slowly, are more inert in terms of temperature, and act as heat storage or buffers. Because of the bigger share of land in the northern hemisphere, it gets warmer during its summer than the southern hemisphere does. The Hadley circulation does not become symmetrical around the equator, but is displaced towards the north. The average position of the TCZ is around 5°N and is displaced much further to the north during the summer than to the south during the winter. This means that the geographical equator and the 'heat' equator do not coincide.

This is clearly shown in the colour plate 'Climate V–VI'. The sawtooth line is the TCZ. During the winter, which is shown in the upper illustration, TCZ is slightly north of the equator except over the Indian Ocean. During the summer TCZ is displaced some distance towards the north and extends in a wide arc up over the South Sahara, over the Arabian peninsula to the Himalayas. The trade winds cross the equator in the Indian Ocean and change into the summer monsoon over Asia.

Most of the rain in the tropics is associated with the cumulo-nimbus clouds of the TCZ. Places along the most northerly position of TCZ only get rain during the summer, i.e. the rainy season. Other places, half-way between the equator and the most northerly position of TCZ get two rainy seasons, one during the spring when TCZ is moving northward and another during the autumn when TCZ is retreating southward again.

The steady winds in the trade wind belts were the motive source for the old-time ships sailing between Europe and North and South America. When they passed the equator they also had to pass through the TCZ. They encountered light winds, and hot and muggy weather, now and then interrupted by an intense thunderstorm. Sailors used to call this region the *doldrums*. Light winds or calms could also be encountered in the subtropical highs and this belt of calm winds was called the *horse latitudes*. The name possibly derives from the times when ships carried horses or cattle across the Atlantic. If the ship ran into very light winds they might be becalmed for such a long time that they ran out of fodder and had to slaughter the horses.

The Rossby Circulation

North of 30°N up to about 70°N, westerly winds predominate in the whole troposphere. The winds and the weather, however, vary considerably from day to day and from season to season. The motion of the air in the belt of westerlies is completely different from the usually steady flow of the tropics, but even if the variations are considerable it is still possible to discern patterns in the air flow.

We have seen earlier that the jet streams and the flow at, for example, 500 mb, show a marked wavy character around the globe. Some of these waves always seem to be present while others appear and move from the west towards the east. The jet streams are the result of the big temperature contrasts between the south and the north parts of the west wind belt. As is shown in the colour plate 'The General Circulation I–II' the net result of the long wavy pattern will be that warm air is transported towards the north in the north-going branches and cold air towards the south in the south-going branches. We can also think of this as a gigantic stirring of the sea of air, heating the northerly latitudes and cooling the southerly ones (just as stirring in a bathtub mixes hot and cold water). This type of circulation is called the *Rossby Circulation* (see Chapter 6).

Why are these waves responsible for the mixing occurring?

In order to see what happens in the atmosphere one can try to illustrate the most fundamental features of the atmosphere in a revolving tank filled with water (or a 'dishpan'). By heating the outer rim and cooling the bottom in the centre of the tank and at the same time varying the speed of rotation it is possible, as already said, to imitate many aspects of atmospheric motion on a rotating earth. In these 'dishpan' experiments it is found that at a moderate rate of rotation and a strong temperature contrast between the edge and the centre of the pan a straight, smooth flow from west to east is obtained. If the rotation is increased a stage is reached when the smooth circular flow suddenly buckles and waves appear, giving a wavy pattern similar to that observed on weather maps.

Meteorologists have known for a long time that small waves can be 'unstable', i.e. grow in strength, intensity or wind speed. We have seen that the polar front cyclones were wave protuberances on a polar front which could very rapidly grow into large-scale depressions. The key to the growth was to be found in the sharp temperature contrasts between hot tropical air and cold polar air concentrated at the polar front. The upper air polar front jet stream is only a reflection of this north–south temperature contrast. We can conclude that temperature contrasts cannot be infinitely strong. If they grow too pronounced the atmosphere releases its energy into cyclones, having the effect of mixing warm and cold air and thereby diminishing the temperature contrast. The atmosphere is in a way a self-regulating system. The atmosphere always appears to find the most efficient way of transporting heat from the equatorial regions to the poles. In the tropics the heat transport takes place by means of a direct thermal circulation – a Hadley Circulation. In the temperate zones (the middle latitudes) the transport is carried out by means of unstable waves – Rossby waves and polar front cyclones that mix the air, thus reducing the temperature contrasts.

The upper right-hand illustration of the colour plate 'The General Circulation III–IV' shows a 500-mb map for March 24, 1970. The picture is fairly complicated, but with some good will it is possible to distinguish four, perhaps five, long waves around the earth. The flow is complicated, however, by a number of polar front cyclones, in different stages of development, moving through the long Rossby waves. One can, for example, see that warm air has streamed far to the north over the Norwegian Sea while cold air masses have penetrated far to the south over eastern North America behind a cold front. In the associated satellite picture below the map the cloud masses of the polar front systems can clearly be seen. The exchange of heat between northern and southern latitudes takes place unceasingly.

Effect of mountains, land and sea

In the northern hemisphere are located some of the highest mountain ranges in the world. The Rocky Mountains in the west of North America extend up to 3–4 km above sea level, thus creating a mighty barrier to the west winds. The high peaks of the Himalayas reach almost through the entire troposphere. In Europe the Alps form a dividing wall between south and north Europe; however, they are stretched out along the mean flow and have less influence on the large-scale motion of the atmosphere. The Rocky Mountains and the Himalayas are big obstacles to the motion of the air since it is either forced to flow around a mountain or to move over it, and often both events take place. We have already seen that on the windward side of a mountain cloudiness and precipitation are increased because of the forced ascent and the condensation of the water vapour in the air. For example, the western slopes of the Scandinavian Mountains get two to three times as much rain as the rest of Scandinavia.

Cherrapunji, on the southern slopes of the Himalayas, for a long time had the world record in precipitation per year – 11,633 mm – as compared with a country such as Sweden where the annual amounts range between 500 to 1600 mm! The mountain ranges also have a far more important effect on the atmosphere. As shown in Fig. 32, the air has to shrink in the vertical when forced upwards, so that an air parcel becomes flattened. The effect will be the same as for the ice skater in the colour plate 'Weather Prediction Using Computers'. The air tends to move as though around an anti-cyclone and over the mountain a high pressure ridge is formed. Further downstream the air starts sinking again; it is stretched in the vertical and the air parcel becomes 'slimmer'. Air moves inward instead of outward and tends to move in a more

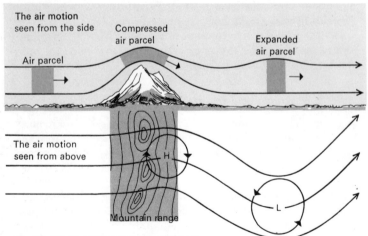

Fig. 32 The flow of air over a mountain range.

cyclonic fashion and a low pressure trough is formed in the lee of the mountain range.

On maps showing the average air flow, and in some of the daily maps, this movement to the south-east of the big mountain ranges is quite evident. The air makes a short visit to the south-east of the Rocky Mountains and then streams in a wide bow up over the North Atlantic and in over central and north Europe. By 'locking' the air current in the long wave extending from North America to Europe the Rocky Mountains are a contributing factor to Europe's comparatively mild climate. This is especially true for the Scandinavian countries. We can thus conclude that mountains play a deciding role on climate in the northern hemisphere by a sort of remote action.

The distribution of land and sea and the difference between these in their capacity to store heat is the other important factor in determining the air flow patterns of the atmosphere. The seas change their temperatures only slowly while the continents rapidly adjust to the changing seasons. During the winter the seas are warm as compared with the cooled continents. The air moving out over the Atlantic from the North American continent is cold at first, but gets heated when it passes over the warm sea surface. This warming has a tendency to decrease the air pressure in lower layers (warm air is lighter than cold), but increase it higher up. Over the Atlantic the distribution of land and sea co-operate with the Rocky Mountains in fixing the air flow in the way described above. In a similar way the atmospheric flow is shaped over eastern Asia and the western parts of the Pacific Ocean.

In the southern hemisphere there are no land masses or mountain ranges within the west wind belt. The westerly winds do not show the same kind of regular deviations towards the north and south as in the northern hemisphere. The number of long waves around the globe is greater in the southern hemisphere, in general between four and six. The westerlies are stronger and polar front cyclones are formed fairly evenly distributed around the globe. The two maps showing the pressure distribution over the northern and southern hemisphere in winter and summer in the colour plate 'Climate V–VI' clearly illustrates this fact. The mean isobars in the southern hemisphere, within the westerlies, extend uninterrupted around the globe in summer as well as in winter. The intense storms moving in the westerlies in the southern hemisphere have given rise to expressions like 'the roaring forties' and 'the screaming fifties', memories from the time when sailing ships rounded Cape Horn and the Cape of Good Hope. In the northern hemisphere on the other hand the mean pressure pattern is dominated by two marked lows, one in the neighbourhood of Iceland, another north-west of Japan. South of these lows there are strong westerly winds. During the summer the lows weaken and the Azores High extends over western Europe. Over the continents, lows have formed in accordance with what we have just seen on the effect of heating at the surface.

Despite the strong influence of mountains and oceans, the air flow can show very large variations. Over North America, however, polar front cyclones are usually formed further to the south while the general northerly drift of the westerlies over the Atlantic causes the polar front cyclones over the eastern Atlantic to take a more northerly track. A reminder of this is the fact that New York, which is at the same latitude as Madrid, has a climate which more closely resembles that of northern Europe than Spain.

The Monsoon circulation

During the summer the whole Asian continent is heated considerably while during the winter it is subject to strong cooling, especially north of the Himalayas. This is also illustrated in the colour plate 'Climate V–VI'. During winter, when air is cooled in lower layers, it gets heavier and sinks because of its greater weight. A large, cold, high pressure zone is formed over all of Asia. During summer the reverse happens. The land masses are heated strongly and pressure falls, so that a comprehensive 'heat low' is formed. In the same way as with sea and land breezes the strong heating during the summer creates a wind towards land, but on a bigger scale. During the spring and early summer the temperature contrast between the Asian continent and the surrounding seas increases, and at the beginning of June the cooler and more humid winds from the Indian Ocean penetrate over India: the start of the summer or south-west monsoon. From this time to the middle of September South Asia gets most of its precipitation. The south-westerly winds over the southern slopes of the Himalayas give a greatly increased rainfall.

In September the Asian continent starts to cool and the south-west monsoon dies away. During the winter there is instead an outflow of colder and drier air towards south and south-west, mainly affecting South Asia. This north-east Monsoon brings dry weather, except in the mountains in South-east Asia where rain occurs during the winter.

The south-west Monsoon is an extension of the south-easterly trade winds of the southern hemisphere into the northern hemisphere, as shown in the colour plate 'Climate V–VI'. This air, which is very moist and hot, also affects the eastern parts of Africa and the South Arabian peninsula. During the winter the two trade winds are recreated and the TCZ which has been situated over South Asia moves down to a position south of the equator. Similar circulations, like the winter and summer Monsoon over Asia, also take place over South America and South Africa during the summer of the southern hemisphere, but on a much smaller scale.

Ocean currents and the hydrological cycle

The global wind systems are also the source of energy for the ocean currents. The Equatorial Stream, shown in the colour plate 'Climate I–II', comes about because the trade winds are communicating their momentum to the sea by means of surface friction. The Gulf Stream, so important to Europe, is forced by the clockwise motion of air around the Azores High over the northern Atlantic becoming what is more correctly called the North Atlantic Drift. The ocean currents contribute substantially to the transport of heat from southern latitudes to northern ones. Computations show that the ocean currents are responsible for about one third of the total poleward heat transport. The connection between the currents of the sea and the atmosphere is not, however, as simple as this. The distribution of land and sea and the temperature variations between different parts of the oceans affect the motion of the air in its turn, once again changing the ocean currents.

The seas are also the source of most of the water vapour in the atmosphere. The seas and the atmosphere are directly coupled to each other through the continuous exchange of water taking place between the two of them. Water mainly evaporates over the tropical seas where the surface temperatures are high. Large amounts of heat are stored in the water vapour of the air as latent heat. A major part of the heat transport to northern latitudes is actually carried out by a transport of latent heat. When clouds and rain form, for example, in the polar front cyclones, this latent heat is released, thereby heating the air.

Water participates in something called the hydrological cycle (see colour plate 'The Hydrological Cycle'). Normally the synoptic meteorologist loses interest in the water when it has fallen to the ground, but it is when it has fallen that man can make use of it. In the process of supplying the whole earth with its annual precipitation the water in the atmosphere has to be renewed thirty times a year.

The climate of the earth

The climate of an area is an important determining factor in many activities and industries, both on the land and in the factory. In modern times, with advanced technology applied to everything from aircraft to housebuilding, knowledge about climatic factors is very important in planning and designing various projects. If one wants to build a drainage system one needs to know how much water is normally expected to drain away and the chances of extremely high rates of rainfall. Bridges and tall buildings have to be designed to resist strong winds and the insulation of houses depends on the temperature and wind climate during the winter.

The average person is probably most interested in the climate of other places when it is time to choose a place to spend a holiday and advice for holiday-makers will be given in the last chapter.

Climate data and climatic periods

Temperature, wind and precipitation at a particular place are reflections of its position in the general circulation, modified more or less by local factors. In climatology meteorologists prepare maps showing the mean temperature for the different months, amount of precipitation, cloudiness, fog frequencies and a number of other useful meteorological quantities such as sunshine duration and radiation conditions. A basis for all these computations are the long series of daily records collected from the synoptic observation stations and special climate stations. The various averages are computed from thirty years of data. The climate period which forms the basis for our present description of climate covers the years 1931 to 1960. The previous standard period was 1901 to 1930. Climate usually changes rather slowly, showing up in the slow changes of the computed averages. However, in some cases there may be fairly large changes in climatic records. At stations which have been situated close to expanding big cities or urban areas large, local, changes of the climate may be noticed.

Different types of climates

In order to characterize more easily the weather conditions at a particular place meteorologists have tried to design different schemes for classifying different climates, for example, by comparing the average temperatures and annual precipitation, or by describing their effects on vegetation. The most common classification is due to the German climatologist W. Köppen who in 1918 formulated a comprehensive, but yet relatively simple classification of climates. He took into account the type of vegetation as well as typical temperatures and amounts of precipitation for a certain place and how these vary between winter and summer. The presentation of world climates in the colour plate 'Climate I–II' is made according to Köppen's classification. But before studying in detail the climate zones one should look at the colour plate 'Climate III–IV' which shows the mean temperatures and precipitation during January and July. From these maps we can get a picture of how the general circulation affects conditions in different places.

The temperature and precipitation climate of the tropics

Starting with the tropics one can see that the mean temperature in general is between 20° and 30°C. The highest mean temperatures are reached in North Africa during July, but large areas in the Asian continent also have high temperatures during this month. In Asia the temperatures during the winter are low while in the tropics they are fairly constant over the year.

The amount of rain during winter has a marked maximum in a zone extending from South America to the central parts of Africa and further on to Indonesia. This band represents the southerly position of the tropical convergence zone. During July, however, the rain maximum is over central America and Africa across to South Asia, where the summer Monsoon gives plenty of rain. The mean rainfall amounts exceed 400 mm in July at many places along the slopes of the Himalayas and in Thailand and Cambodia. The TCZ over Africa and central America can give between 100 and 400 mm during the summer. The rain comes to an end in the southern Sahara where there is less than 25 mm a year; some desert places never get any rain. Similar dry areas are found in Mongolia, the Arabian peninsula, the inner parts of Australia and east of the Rocky Mountains in U.S.A. We can also notice another systematic feature in the distribution of precipitation. The farther inland we go over the continents the drier the climate becomes as well during winter as in summer (except in the tropics).

In the subtropical regions, immediately north of the tropics, we find the really dry desert areas. These become very hot during the summer, but fairly cool during winter when the temperature can go well below freezing point during the nights because of outgoing heat radiation. Vegetation is sparse. One exception is again Southeast Asia which is under the influence of the Monsoon circulation.

The climate of Britain

The weather of the middle latitudes is determined by their position in the west wind belt. The western parts of the continents have relatively mild winters, but cool summers because of the influence from the sea. A good example of such a maritime climate is that of Britain.

Although the British Isles are only very small in area compared to the rest of the northern hemisphere, they contain a surprisingly wide variety of climates, ranging from the almost semi-arid conditions in East Anglia, to the nearly subtropical climates of Cornwall and Kerry, and to the areas of the Scottish Highlands which are akin to an Arctic tundra.

The reason for this is that Britain lies on the border between the wide Atlantic to the west and the large land mass of Europe and Asia to the east, and furthermore it is almost half-way between the cold polar regions to the north and the warm tropics to the south. So much then depends, not only on the seasonal strength of the sun, but also on the direction of the wind. A southerly wind in mid-winter can result in higher temperatures than a northerly wind in mid-summer.

Conditions are so variable, from day to day and season to season, that it has been said with some justification that 'it does not have a climate, it only has weather'. Iceland also has a similar saying 'if you don't like our weather, then wait five minutes'. It must be stressed, however, that these variations are rarely carried to excess, and Britain does not suffer from the extremes of weather that can be experienced in other parts of the world.

Nevertheless, despite these ceaseless variations, some sort of general pattern does emerge, even though this pattern is oversimplified by the simple meteorological averages. The west is different to the east, the north to the south; it is certainly different in farming patterns, in general living conditions, and perhaps even in the effects shown in human character and behaviour patterns.

To begin with temperature, the Atlantic Ocean ensures that the western coasts enjoy a more temperate climate, being cooler in summer, and milder, sometimes much milder, in winter. The North Sea to the east, being smaller in extent has a similar, but much less pronounced effect on the eastern coasts. The south-east of England, being closest to the continent of Europe, is liable to have the warmest summers and, in times of frequent southeasterly or easterly winds, by far the most severe type of winter weather. Temperatures also tend, on average, to decrease with height, at the rate of about 0·6°C for every 100 m, a feature of mean values which must not be confused with the fact that the lowest minimum temperatures on a night of radiation frost are to be found in the river valleys.

There is a wide range of rainfall climates in Britain, because rain increases with proximity to the Atlantic and with increase in height above sea level; both these effects are combined because the highest ground lies to the west and north. Average annual rainfalls can range from under 500 mm in the driest parts of East Anglia to almost ten times this amount in the Lake District of the north-west. The annual rainfall is, on average, fairly evenly distributed through the months of the year, and it is only in the hills and the extreme south-west where winter rainfall is dominant; in some parts of central England there is, on average, more rain in summer than in winter. There is, however, a big year-to-year variation in monthly rainfall totals about the long-term average, and within a decade it can be as little as one-fifth, or as great as twice the mean figure.

Both Scotland and Ireland show similar patterns of average rainfall, being least in the lowlands of the south-east and highest in the west. These countries are also less liable to droughts than are England or Wales.

Sunshine in Britain shows a marked seasonal variation. This is because there is a big change in day length from about eighteen hours in the extreme south in mid-summer to less than six hours in the far north in mid-winter. Winter also the more cloudy period so that in December the average duration of bright sunshine is about one hour, but in high summer it can be seven or eight times greater.

The areas with most sunshine lie close to the coast, especially the south coast, and are at a low height above sea level; sunshine is least at places well inland and on the hills and mountains. The pattern is slightly different in Ireland which enjoys most sunshine in the south-east and least in the north-west.

With the exception of western Norway, Britain is probably the windiest part of Europe. Average annual wind speeds range from 3–3·5 m/sec. in inland areas of south-east England, to over twice this amount on the western coasts. Gales are mainly experienced in the winter half-year, but summer gales are not unknown on the north-western coasts.

England's reputation for fog is probably overstated, being a relic of the heavy industrial pollution of the last century. With the effective implementation of Clean Air Acts since World War II, the fogs, mainly in autumn and winter, are far less dense, but even so they can cause very dangerous conditions for traffic, especially on the motorways.

Even a moderate fall of snow in winter seems to have a major effect on traffic systems. This is partly because such snow is not a regular occurrence, and so it is not worth while keeping expensive snow clearance equipment in readiness. In Scandinavian countries, with a regular winter snowfall, far more efficient arrangements can be economically made.

Thunderstorms, mainly a summer phenomena, are not unduly frequent; they are most likely to occur in the south-eastern parts of England, and are relatively rare in Scotland.

Energy sources

Before the days of the widespread use of coal and electricity, many rural industries were powered by windmills or water mills; most have now disappeared. Now, with the threat of an energy crisis becoming more and more apparent, attention is again turning to squrces of 'free' power.

There is little doubt that wind, harnessed extensively at individual sites could only supplement, and never replace our increasing energy needs. Solar panels in house roofs could also supply some energy, but only to any appreciable degree in summer, when the demand for domestic use is least. It is a pity that the public arguments (usually boiling down to for or against nuclear power), are often based on that continually varying and utterly undependable unit, namely money. The time may yet arrive when it will be economic to supplement our energy from natural sources, but at present it is certain that the best action to take is one involving insulation, to avoid the loss of heat from buildings, which is most important in a country which is subject to more winds than most. At least the British Isles do not have to face the American problem of using expensive fuels to keep themselves cool in summer.

Maritime and Continental climates

The farther to the east we go over the European continent, the bigger the temperature differences between summer and winter.

The inner parts of the U.S.S.R. have very cold winters with mean temperatures between $-10°$ and $-20°C$ in January; during the summer the temperature is about $+10°$ to $+20°C$. The temperature amplitude during the year is therefore large in continental climates.

The maritime climates, influenced by the sea, have rain all the year round. Over the continents most of the precipitation falls during the summer as showers. In many places around the earth there is a strong orographic enhancement of precipitation, shown clearly on the precipitation maps. The Rocky Mountains, the Scandinavian Mountain range and the Alps get more rain than their surroundings. In the southern hemisphere only the southernmost tip of the Andes is within the west wind belt and gets large amounts of precipitation all the year round. Over the polar cap there are predominantly easterly winds. The temperature is low the whole year and the air dry, so that there is little precipitation. Sometimes, however, polar front cyclones or lows formed along the arctic front can give profuse snowfalls that tend to accumulate with little opportunity for melting. The southernmost part of Greenland is, however, affected by the migrating Atlantic storms and gets a lot of precipitation all the year round.

The climatic zones

The plate 'Climate I–II' shows the climates of earth according to Köppen's classification; the key describes the different climate types. Köppen's classification is based on the combined variation of temperature and precipitation. The arid climates are, however, only characterized by the amount of rain while the polar climates are classified according to temperature. The rainy, tropical climates are found around the equator. The mean temperature is above $+18°C$ all the year and varies little between the seasons. The climate of the rain forests is found in the regions which are affected by TCZ all the year while the tropical semi-arid regions only have one rainy season. Further to the north, or south, the tropical climate changes into the dry climates of the steppes and the deserts. Dry climates can also be found outside the tropic of

Cancer in the inner parts of western Asia, inner Australia and in the western parts of North and South America.

A special kind of climate is the Mediterranean climate. It is typified by rain usually falling during winter while the summers are dry. The Mediterranean moderates the temperatures during the year making the winters fairly mild while summers do not get terribly hot. The southerly track of polar front cyclones during winter give the Mediterranean area most of its rain. An exception is the inner parts of Spain where the surrounding mountain ranges create a dry and continental type of climate. The mean temperature of the hottest month is above $+20°$ while the coldest is above $-3°C$. Winter temperatures can at times reach fairly low values in the northern Mediterranean countries. The whole of Central Europe is on the other hand characterized by a warm temperate, humid, climate with rain or snow all the year round. This is also true for southern Scandinavia which is under a maritime influence keeping the temperature during winter on the average above $-3°C$. A similar type of climate is also found in south-eastern U.S.A. The level of $-3°C$ is chosen because it indicates the limit for a continuous snow cover during part of the winter, even if a particular winter happens to be milder than normal. A large part of Scandinavia has a cold temperate climate with the coldest month below $-3°C$ indicating a continuous snow cover during the winter. Northern Sweden is comparatively dry while southern Sweden has rain all the year round. Practically all of Siberia and Canada belong to the cold temperate climate with dry winters. Polar climate with all months below $-3°C$ can either be glacial or of the tundra type. The glacial climate persists in the inner parts of Greenland and over the whole polar cap. A polar climate is also found at higher altitudes in the mountain areas.

It is also interesting to look at the ocean currents which play such an important role in the creation of the different climates. In the northern hemisphere one can distinguish two very marked and extensive bands of streaming water in which water is transported towards the north. In the Atlantic Ocean, the Gulf Stream and, in the Pacific, the Kuroshio Current, act like giant oceanic anti-cyclones. Also in the oceans something resembling atmospheric fronts may form. East of Newfoundland the Labrador Current carries very cold water in close contact with the warm waters of the Gulf Stream and in this area there is a very sharp temperature difference between north and south. The Gulf Stream meanders along the Atlantic Coast of North America. The temperature along the shores can change very rapidly as anyone who has gone swimming along the beaches of, for example, Long Island, can testify. Warm air streaming up from the south or south-west is cooled very rapidly when it passes out over the cold water of the Labrador Current. Fogs easily form and the waters outside New-foundland have the highest frequency of fog in the world. West of the South and North American continents there are two cold currents, the Humboldt or Peruvian Current and the California Current. The Humboldt Current brings cold water rich in nutrients towards the equator and is very important to fishing. The Humboldt Current has made Peru the biggest fishing nation in the world. The California Current brings cold water down along the California coast giving rise to persistent fogs or low stratus clouds over the sea. The colour plate 'Fog' shows how the fog rolls in from the sea wrapping the Golden Gate Bridge in white clouds. The really warm ocean surfaces are found in the Equatorial Currents, in the Caribbean Sea and the Persian Gulf.

The saw-tooth line in colour plate 'Climate I–II' denotes the southernmost mean position of drifting ice during winter. In the North Atlantic icebergs can reach as far south as Iceland.

World Weather Records

To finish this short summary of world climates here is a selection of world weather records. The 'highest and the lowest' gives a feeling for the tremendous variation of weather around the world as shown in the following statistics:

Air pressure at sea level
Highest: 1083·8 mb recorded in Agata in U.S.S.R., December 31, 1968
Lowest: 874·0 mb recorded in typhoon Ida in the Pacific Ocean

Wind speed
Greatest wind gust at the surface: 116 m/sec.

Five minute average wind speed: 94 m/sec.
Both recorded April 12, 1934 at Mt. Washington, New Hampshire, U.S.A.

Temperature
Highest temperature (in the shade): $+58°$ in El Azizia, Libya
Lowest temperature: $-88·3°C$ in Vostok, Antarctic
Highest annual mean temperature: $+34·4°C$ in Dallol, Ethiopia
Highest sea surface temperature: $+35·6°C$ in the Persian Gulf

Precipitation
World's greatest amount of rain in one minute: 31·2 mm in Unionville, Maryland, U.S.A., July 4, 1956
Greatest amount in twenty minutes: 206 mm in Curtea de Arges in Romania
Greatest amount in twenty-four hours: 1245 mm in Paishih, Taiwan
Greatest amount in one month: 9300 mm in Cherrapunji, India, July 1861
Greatest mean annual precipitation: 11,684 mm (1912–45), Hawaii

Climate change

We must remember that world climate is continually changing, and the first forty years of this century were probably the warmest in north-west Europe for many centuries. There is now great awareness of the possibility of climatic change, but a great division of expert opinion on the direction it will take. One body of thought postulates a return to a 'little ice age', another opines that the increasing amounts of carbon dioxide in the atmosphere will lead to further warming. Time alone will tell us the answer, because of all forms of weather forecasting, that concerning climatic change presents the most difficulties.

Local climates

That weather and climate can be different in Stockholm and London is not surprising to most people, but that there can be a large difference between Glasgow and Edinburgh might be somewhat more amazing. However, it is none the less true that large variations can occur over fairly short distances in temperature, wind and rainfall. These small-scale variations are due to local conditions such as the different distribution of the surrounding fields, forests, buildings, etc., and the neighbourhood of hills, plains, valleys, lakes and rivers and seas which change the climatic conditions.

Urban climate

The climate and weather conditions of big cities differ considerably from those of the surrounding rural areas. Colour plate 'Local Climates and Air Pollution' shows what happens over a big city in summer and winter.

The air pollution, which forms in the city, can be caught beneath a dome-shaped inversion. During winter, when the air gets trapped and is not transported away, windless conditions can persist for long periods. Such periods are usually cold and more oil and coal have to be burnt to heat the houses. The air pollution level rises rapidly and the cities become unhealthy to live or work in.

Hills and valleys

Weather conditions can also be very different between hills and valleys. During winter and during the night the cold air flows down into the bottoms of the valleys because of its greater weight. The temperature at night becomes lower in the valleys than on the surrounding hills. The generally lower valley temperatures increase the frequency of fog and the risks of frost. The winds are weaker in the valleys than on the hills and in long valleys the wind can only blow along the direction of the valley. During the summer the valley gets higher daytime temperatures than the hills. Places in the lee of higher hills or mountains get less rain than places on the windward side.

Forests and plains

Temperature is usually higher in forest areas than out over a plain in winter while the reverse is true during summer. This is because the plains react quicker to incoming and outgoing radiation than the forests, which can store large amounts of heat in the trees. The increased roughness of the trees makes the surface winds weaker than over the comparatively smooth plains. Trees and vegetation can store large amounts of water and the deep root systems of the trees can extract the ground water which can be given off to the atmosphere. Humidity increases in woods while the strong evaporation over the plains easily dries out the soil. Trees have been used for ages in order to change the local climate. The planting of forests or trees around farms give weaker winds and higher daytime temperatures. Trees also decrease the risk of wind erosion, and so help the growth of crops. In many places on earth the loss of soil by wind erosion is a serious problem, but it can, however, be reduced by the planting of shelter-belts and windbreaks.

Lakes and rivers

The proximity to water has a marked influence on climate. Large lakes and seas have a moderating effect on the temperature during the course of the year, giving milder winters and cooler summers than places far removed from water. Inland waters, however, tend to increase the frequency of fogs.

It is not only the natural features which affect the local climate. The urban areas have a climate of their own and other encroachments on 'mother nature' also occur. Warm waste water released into lakes and coastal waters leads to changed temperature conditions in the water and in the air above.

Air pollution

By the increased use of fossil fuels, i.e. oil, petrol, paraffin and coal, for heating and energy production we release more and more pollution into the atmosphere. The oils we burn contain sulphur and, besides water and carbon dioxide which are not poisonous in themselves, carbon monoxide and sulphur dioxide are released into the air as products of combustion. In many industrial processes dust and various chemicals are produced and released into the atmosphere or into lakes and seas.

Sulphur dioxide is probably our commonest air pollutant. In the air and particularly in the clouds sulphur dioxide gradually gets transformed into sulphuric acid, a highly corrosive acid. Half of all the sulphuric acid in the rain over Sweden comes from west and central Europe, the south-western parts being particularly exposed to this form of pollution. Sulphur dioxide can be transported quite a long distance before getting washed out by the rain. The acid-rain falls into lakes and soils gradually increasing the acidity in the ground and in the water. At some places in west Sweden the acidity has become so high that lakes have become completely devoid of life (see the lower part of colour plate 'Local Climates and Air Pollution').

Several million tonnes of carbon monoxide gas are released annually by motor traffic. This is a poisonous, odourless and colourless gas, but after some time, is transformed into carbon dioxide. In the fumes from a car exhaust there are also oxides of nitrogen and hydro-carbons. Under the influence of the ultra-violet radiation from the sun, so-called photochemical smog may form. This happens in large urban areas with heavy car traffic such as Los Angeles, Tokyo and New York where the air can settle for a long time without being carried away. The chemicals in the photochemical smog are corrosive and harmful to man; smog adversely affects the vegetation and causes eye irritation. In extreme cases and in combination with strong heat the situation becomes unbearable so that it is practically impossible to be out-doors.

During the twentieth century the development of more and more waste-producing technologies has taken place virtually un-checked. During the 1960s a necessary, but unpleasant awareness of this arose and since then deep concern has been growing about the danger and imminent threat of the increased amounts of air pollutants, but we are still releasing enormous amounts of chemicals, dust and particles of various kinds into the atmosphere. However, it is now becoming more difficult to get permission for the building of new industrial or power plants without taking action to reduce or control the pollutants emitted into the air.

Many countries have introduced legislation regarding the limits for the acceptable levels of concentration in the air of many compounds, in order to let people live a fairly normal life. The problems are worst in heavily industrialized parts of the U.S.A., England and West Germany, where the environment can at times be very unpleasant to live in.

The atmosphere has a self-cleaning capability which gradually breaks down many substances into harmless compounds or washes them out by rain. This is especially true of substances which occur naturally in the air.

A natural goal for an environmental protection policy would be not to allow the release of more pollution into the atmosphere than what it can take care of by its own natural self-cleaning mechanisms. This policy is far from being applied today. Instead some countries apply the principle of as little dirt on their own territory as possible – and let someone else have it! Air pollution can be transported over very long distances as many investigations have shown, even all around the globe. What actually happens to the pollutants after they have been emitted into the air is a question for meteorologists to study. How much is released is a political matter. Meteorologists and other scientists work on how air pollutants react chemically in the air, how they are dispersed, transported, and diluted, and how they are removed in various ways from the atmosphere, as for example by being washed out by rain. The results of such studies can be applied directly to planning new industries, in designing high smoke stacks and the composition of the smoke released from the chimneys. The immediate goal is to keep the concentrations of various substances below a level which would cause harm to people living in the zone of influence.

A general rule is that the higher a smoke stack is built the more smoke can be released without harmful risks to the immediate environment. It is also more profitable to have a big smoke stack than many small chimneys as shown in Fig. 33.

If the emissions are not reduced the effect of a chimney will only be that the air pollutants will reach the ground further away. If a highly polluting smoke stack is close to the border of another country that receives the contaminating air it emits international complications might easily develop.

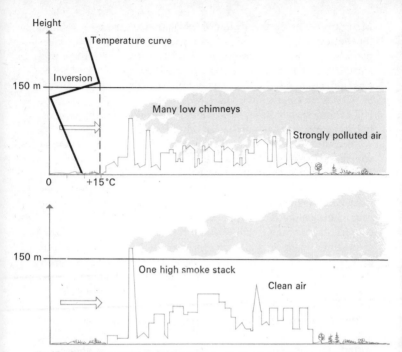

Fig. 33 Over urban areas ground inversions, or elevated inversions, often occur, especially during winter. The occurrence of inversions makes it more advantageous to build fewer but larger power plants and factories with high smoke stacks than many small ones. Many small, low chimneys and smoke stacks emit their smoke beneath the inversion where it accumulates while the smoke from high smoke stacks is carried away by the wind without reaching the ground.

It is clear that the general level of release of air pollutants must be reduced sooner or later. Meanwhile meteorologists can contribute an improvement in the situation in many ways. By studying how quickly pollutants are transported away from different areas – the so-called ventilation climate – local governments are provided with a firmer basis for making decisions on where to establish new industries. Maps concerning the ventilation climate have been worked out for many countries. It is also well known that the most severe cases of air pollution occur in special weather situations, often in anticyclonic situations with weak winds and poor ventilation. The meteorologist can warn about impending severe air pollution situations making it possible for the undertaking of counter measures such as reducing car traffic and smoke stack emissions. In many big cities in the U.S.A. the air pollution level is monitored continuously every day and a special air pollution index is given in the daily weather reports. When the index gets too high people with weak resistance, with heart diseases and respiratory troubles, etc., are advised to stay indoors.

So far we have only discussed a small sample of all the substances emitted into the atmosphere. Many of these occur naturally in the atmosphere; others do not exist naturally, but are transformed into naturally occurring ones.

In December 1952 London suffered a pollution catastrophe. In little more than a week 4000 people died when about 500,000 tons of sulphur dioxide hung in a thick cloud over London; the dust from all the thousands of coal-burning fires caused the formation of a thick smog. This classical 'London fog' is nowadays fairly rare. After the serious pollution in 1953 a big programme to limit the emissions was launched. The main source of pollution was the numerous coal-heated private houses, which, through an incomplete combustion, released large amounts of coal dust and sulphur dioxide into the air.

Coal as a source of air pollution is now under control in London, but all cities suffer from the contaminants emitted by millions of cars, made worse by the presence of poisonous lead compounds.

Carbon dioxide

Carbon dioxide occurs naturally in the atmosphere by an amount of about 0·03% (or 300 parts per million). Carbon monoxide and carbon dioxide are produced by combustion, and the air we exhale contains carbon dioxide which the plants use in the photosynthesis cycle. Carbon dioxide achieves a prominent role by its large effect on long-wave, out-going, heat radiation from the earth. As we have seen earlier, carbon dioxide and water vapour were responsible for the so-called greenhouse effect causing the average temperature of the earth to become more than 20°C higher than it otherwise would be. Since the beginning of this century the content of carbon dioxide in the atmosphere has increased about 10% due to the increase in combustion of oils and coal. If we continue to release carbon dioxide at the same rate as now the amount of carbon dioxide in the year 2000 could be 360 ppm or 20% more than in the year 1900. What sort of consequences could such an increase of carbon dioxide result in? Calculations in the U.S.A., where scientists have tried to simulate the effect of an increase in the carbon dioxide content by very complicated computer models, seem to indicate a general warming up of the atmosphere, such that if the content of carbon dioxide in the air was doubled, the temperature could increase about 4° on average over the globe! Simulations of this kind should, however, at the present time be regarded with some scepticism since only the effect of an increase in carbon dioxide has been considered. Computations in retrospect show at least that carbon dioxide might be responsible for a general temperature increase of only a half to one degree since the beginning of this century.

Climatic changes and man's impact on climate

The example of carbon dioxide shows that it is possible for man to affect and change the climate on earth. In the beginning of the 1950s the possibility was discussed that hydrogen bomb tests were responsible for the bad summers of the beginning of the 1950s as compared with previous years, but meteorologists were not convinced. The hydrogen bomb tests have probably had no effect at all on climate, but the thought that man himself might be responsible for deterioration in climate is not new even if the possibility of man affecting the climate on a global scale is of very recent date. Local climatic changes have often been brought about by the actions of man, desertification being a case in point.

Past climates

Going back in time one notices that climate must have changed considerably many times through the ages. Billions of years ago these variations and changes were probably due to the physical development of the planet Earth as it cooled down and the composition of the present atmosphere stabilized. For other drastic and revolutionary changes it is more difficult to find satisfactory explanations. Before it is possible to draw any conclusion about the future, however, one has to understand the past and be able to explain the climatic changes which have already occurred. How then has the climate changed?

By studying the layered structure of the thick ice sheet covering Greenland by means of deep drilling into the ice, it is possible to deduce how temperature and precipitation have varied through many thousands of years. Tree rings and texts and books preserved from ancient times also serve as valuable sources for early climates. The history of man is of course closely linked to the climate and its variations. The so-called Harrapan civilization flowered in the Indus Valley (in present Pakistan) 4000 years ago; grain, melons and cotton were grown and the people possessed a highly developed craftsmanship. About 1800 B.C. the Harrapan culture died and the land was invaded by others. Nowadays the area is a sterile desert, but pollen findings have shown that this area was once covered by fertile vegetation. A few years ago, after prolonged droughts in the south Saharan countries, people left the border regions and moved southward. These examples, from the distant past and from the present era show that changes in the climate can have drastic consequences for the people involved.

Figure 34 shows a compilation of the variation of the temperature climate since the last ice age. After A.D. 1880 the values are based on regular temperature measurements over the northern hemisphere (shown in the uppermost diagram). Looking first at the lowest diagram, however, one can see that at about 15,000 years B.C. the last ice age started to draw towards its end. By 10,000 years B.C. the large ice sheet had withdrawn northward and human beings could follow and settle in the formerly ice-covered areas. Temperature has also varied considerably since those times; at the time of Christ the climate was warm while the following seven hundred years were colder than normal. Our last millennium commenced with a very favourable climate over the northern hemisphere, and it was during this time that the Vikings colonized Iceland and Leif Eriksson carried out his great sea voyage to Vinland (North America). This mild period lasted until

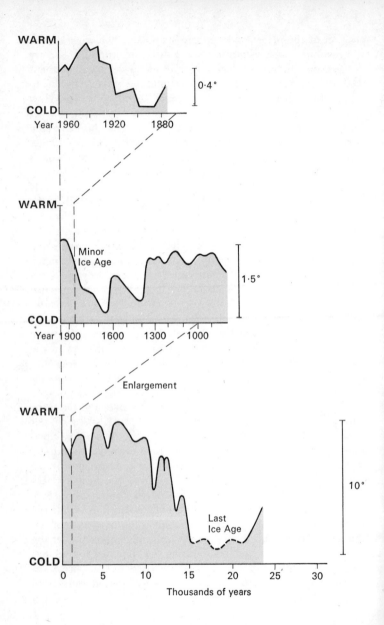

the end of the Middle Ages when the climate suddenly became harsher. After a short peak at the end of the sixteenth century the long cold period, that culminated in the middle of the seventeenth century, began.

This cold period has been given the name 'little ice age' when many glaciers advanced towards inhabited areas. In the Netherlands the canals froze and it snowed far south into the Mediterranean area. During the eighteenth and nineteenth centuries it gradually became warmer again, and our daily temperature records first date from this period. Towards the end of the nineteenth century a small relapse occurred with some very severe winters. During the twentieth century the average annual temperature has increased, by about 0·5°. This warming ceased in the 1940s and since then it has become slightly colder again. Looking at the graph we can see that in the twentieth century we have had the most favourable climate ever over the last 2000 years. During this time man has developed an amazingly high technological and material culture. The warm climate has been favourable to northern agriculture, and made it possible to invest in industrial development. It is, however, very difficult to say if the couple of warm years recently are due to random variations or if they indicate a real trend upward again. Some very recent calculations of the average temperature in the northern hemisphere point in that direction. The 1970s have not only brought a slight warming up, but also brought droughts in many places. In 1972 the U.S.S.R. was hit by a severe drought forcing it to buy grain on the international market. In India the Monsoon has given much less rain than normal. The prolonged drought in the Sahel countries has no counterpart since the beginning of this century. In 1976, there

---

Fig. 34   The variation of climate during 20,000 years.
The three diagrams opposite show the variation in the temperature of the northern hemisphere at different times through the ages. The lowest diagram, which results from probing the ice sheet in the large glaciers, from tree ring data and other sources, shows the great changes in temperature since the last ice age which ended about 10,000 years ago. The middle diagram is a 'blow-up' of the last 1,000 years based on historical notes and old chronicles. The temperature variations during the last 100 years, as shown in the upper diagram are based on real observations from the northern hemisphere. The heat wave during the twentieth century culminated during the 1940s. The bars on the right of the diagrams show the limits within which the temperature varied during the different periods, e.g. 0·4°C during the last 100 years.

was a serious shortage of summer rain in many countries of north-west Europe.

During the twentieth century the number of severe northern winters have decreased even if the very cold winters in the 1940s, and in 1962, seem to contradict this. The recent computations of northern hemispheric mean temperatures on the other hand indicate that the range of variations from year to year has increased, giving a greater tendency to extreme conditions in many places.

A statistical 'forecast' of climate

From a study of the variation of mean temperature of the northern hemisphere it seems possible to discern certain periodic rhythms with changes between warm and cold. These periods can be analysed mathematically. If we assume that past periodicities will continue an analysis of this kind can be used to extrapolate the temperature curve in Fig. 34 into the coming hundred years. The mathematical formulas in this case predict a general cooling off of the northern hemisphere. Considering the coming and going of the ice ages it furthermore looks like time for another ice age in one or the other of the coming millennia. The statistically predicted cooling would, however, only be a short 'intermezzo' and at the turn of the next century we would be on our way up again! A statistical calculation of this kind, however, does not take into account the activities of man himself and what influence they might have on the climate. Perhaps the increased emissions of carbon dioxide may counteract this trend downward? Man's awareness of the effects of the variations in climate on his activities, particularly in the production of food, is an incentive and encouragement for a massive scientific attack on climatological problems. We have got to know and understand how and why climate is changing and what may happen in the future.

Causes of climatic changes

Having emphasized how important it is to understand the causes of previous climatic changes it is embarrassing to have to confess how little we do actually know. In fact, none of earlier great

reversals in climate during the last 20,000 years has been satis-
factorily explained. Instead we have a whole series of possible
causes – or impossible causes.

The most apparent source for variations in the climate is the
sun. It supplies all the earth with energy, a fraction of which goes
into the motion of the atmosphere. If the sun was weaker or
stronger that would immediately give rise to changes in the
climate. The intensity of solar radiation is measured by the so-
called solar constant, i.e. the amount of energy passing through
one square metre at the outer limit of the atmosphere. But the
solar constant – amounting to about 1400 watts – has not changed
as far as one knows for the last 10,000 years. Variations in the
amount of heat received at the surface of the earth can come about
by varying amounts of dust in the atmosphere. An increased
amount of dust would absorb and reflect more of the solar radiation causing a cooling at the surface. At many times volcanoes
have ejected large amounts of dust into the atmosphere, particularly into the stratosphere. At these times a general lowering of the global mean temperature has been detected, though
these changes have only lasted for a couple of years and never
persisted.

'But what about the sun spots then?' the inquisitive reader
might ask. That the energy flows towards earth vary following a
basic eleven-year sun-spot cycle is well known. However, the
resulting energy variations only take place in the ultra-violet range
of the spectrum and for even shorter wave lengths. The energy in
that part of the spectrum only makes up a small fraction of the
total energy and most of it is absorbed above 50 km anyway. The
sun spots do not seem to be a probable explanation for the variations in climate, although many attempts have been made to in-
culpate them.

So far the contributions of man to the atmosphere have been
small except for carbon dioxide. With the increasing amounts
released in the air, however, that might not necessarily be true in
the future. In a number of areas our interference might lead to
substantial changes, and indeed this is already the case judging by
conditions in and around the big cities. Other examples are found
in the increased cloudiness in the upper air over the high traffic

density air routes over Europe and North America, and such clouds may diminish the solar radiation reaching the ground.

The emissions of air pollution change the composition of the air and its properties, especially those concerning radiation. Freons, nitrous oxides and other chemicals can affect the ozone layer with consequences at the surface of the earth. Even if these releases are spread over the whole earth they are still not abundant enough at the present time to cause immediate harm. Attention has been focused on the threats inherent in their presence and legislation against further emission is, so to speak, 'in the air' for a number of compounds.

The interactions between the oceans and the atmosphere are too complicated to make it possible to say anything decisive about the climate and its future behaviour. We have become aware of some risks and intensive scientific work is needed to give us insight into the mechanisms of the climate. The chief aid in this work may be very elaborate and comprehensive computer models of the atmosphere wherein all its components and their interactions are incorporated. Data must be included on the interactions between the oceans, the surface of the earth, the snow cover and vegetation and their influence on long- and short-wave radiation; the chemical reactions in the air have also to be modelled. By means of these 'super' models, being developed at several research institutes around the world, it is possible to conduct 'controlled' experiments. One can study what would happen if the amount of carbon dioxide doubles as previously discussed. One can also see what could happen if the snow-covered areas on earth increased. One would expect that the greater reflectivity of the snow would cause less radiation to remain at the surface. Temperature would fall and more snow would accumulate on the surface giving a further decrease in temperature; such a chain of events is referred to as 'snow instability'. By decreasing the snow cover we would change our climate in a similar, but opposite way, and get a general melting of the ice and a temperature increase. One theory of the ice ages is that they started one winter with exceptionally heavy snowfalls which invoked the snow instability mechanism.

As we have seen, the consequences of even minor displacements of temperature of the order of $\frac{1}{2}$–$1°$ can become very serious. Such changes in the mean temperature would determine in which latitude bands agriculture would be possible. Cyclone tracks and jet streams would be displaced towards the north or south changing the precipitation patterns. Former rainy areas would become dry and dry areas become wet. If we know where and when climatic changes will occur we might perhaps undertake counteraction to mitigate the effects of unfavourable changes. All attempts to change the climate in any way by major manipulations, however, must be banned as long as we know so little about the very complicated interaction between the atmosphere, the oceans, the biosphere and the chemical reactions in the air. When it comes to smaller weather systems or phenomena, though, we no longer need to sit idle. Many meteorological phenomena are nowadays influenced by man's actions.

Weather modification

'Everyone talks about the weather, but no one does anything about it.' These words by Mark Twain are no longer entirely true. Concentrated efforts to change the weather are carried out in many places all over the world. At the present time, however, the possibilities for tampering with the extensive polar front cyclones are very slim. However, when concerned with a number of other, smaller scale, phenomena many people have taken up the challenge, and in other cases the prime motive has been to produce rain.

Hail suppression in the U.S.S.R.

Some weather phenomena cause considerable damage to crops every year; in the U.S.S.R. hail causes great crop losses. Hail is formed in the very strong up- and down-draughts in cumulonimbus clouds when supercooled drops are lifted up and down several times and grow in size by means of collisions with other drops or snow crystals. If it was possible to influence a cumulonimbus cloud at an early stage of its growth it would perhaps be possible to get the cloud to produce rain before entering the hail-producing stage. This could be achieved by injecting freezing nuclei into the cloud, thus getting a quicker transformation into the ice stage and thereby a faster production of rain. Silver iodide has been used extensively as freezing nuclei since it has a suitable crystal structure. Frozen carbon dioxide has also been used. In the U.S.S.R. silver iodide is injected into the cloud by anti-aircraft guns. Shells are loaded with silver iodide and explode in the middle of the cloud as shown in the colour plate 'Weather Modification'. Significant results have been claimed by the Russians.

Project Stormfury

Tropical cyclones are more extensive weather systems which bring hurricane-force winds, rain storms and floods to the coasts of many countries. To be able to control such conditions would be of great value, but it is not possible for man to subdue a tropical cyclone. Even a small reduction, however, of the strongest winds would substantially decrease the damage caused by a tropical cyclone. Thoughts of this kind lay behind a project in the U.S.A. called 'Project Stormfury'. During the 1960s preliminary tests were performed by seeding tropical cyclones with silver iodide (see colour plate 'Weather Modification'). In August 1969 an attempt was made with storm 'Debbie' which was considered to be very encouraging. What seemed to happen was that if silver iodide was spread over the violent cumulonimbus clouds just outside the eye of the storm the ring of precipitation was widened and at the same time the winds beneath at the surface were somewhat reduced. The reduction in wind speed for the 'Debbie' case was between 7–15 m/sec. That meant a decrease from hurricane to gale force. Such a decrease would make quite a difference to the extent of damage. These attempts and experiments were pursued in the Pacific, but discontinued later because of international disagreement on the total effect of these manipulations of the weather.

Clearing of fog

All aircraft need good visibility when landing, and to a somewhat lesser degree when taking off. Fog at an airport is capable, even with the modern electronic landing equipment of our days, of bringing air traffic to a standstill. At the Charles de Gaulle Airport outside Paris attempts have been made to clear the runways from fog. In order to do this fifteen jet engines have been placed along the runway. When the fog rolls in the engines are started and the turbulence and the heat generated is sufficient to clear the fog long enough for aircraft to land. This method is simply an extension of a method used by the British during World War II. In order to aid returning bombers to land safely from their raids over Germany a great number of paraffin-fired flame throwers were mounted along the runway at one airport. When the aircraft were on the way in the burners were turned on and for a few but important minutes the fog was dissolved over a limited area. Some pilots expressed the opinion that the flames were worse than the fog!

When the fog is clearing at the airport after jet engines have been running it is easy to assess the impact of the weather modifying effort. It is, however, not so clear that the shooting of silver iodide grenades into cumulonimbus clouds and the seeding of tropical cyclones really have an effect. What one believes is the result of weather modification might very well have happened anyway. After we have 'treated' a particular object it is impossible to find out what would have happened naturally. These difficulties to control and assess the effect of a certain action have lead to doubts so far as weather modification is concerned. These have been strengthened perhaps by the somewhat ruthless exploitation of people's credulity by certain entrepreneurs. In the American Midwest there are many small private firms operating with aircraft seeding cumulus clouds to stimulate rainfall over the fields of farmers. Whether or not they actually cause rain to fall is very difficult to prove.

More serious experiments have shown, however, that weather modification is feasible and the field is open for anyone to try. Rain is an important event in many places where there is a dry climate. If someone upwind of a location tried to extract the little the clouds have to give, with nothing left for those downwind, serious conflicts could develop. If one country is stealing rain from another disputes are raised to an international level. Serious negotiations have already started to try to reach agreements about the air and what interferences will be accepted.

Weather modification has also been used in warfare. In the end of the 1960s and early 1970s the U.S.A. dropped silver iodide on the clouds over the so-called Ho Chi Minh trail from North Vietnam via Laos and Cambodia to South Vietnam. The intent was to encourage the Monsoon rains making roads muddy and difficult to negotiate so that the transport of supplies to the south by the enemy would be delayed. In this project, research results from Florida and Australia were used. The reports of such meteorological warfare almost caused a minor crisis within the meteorological world. The weather forecasts, as we have seen, are so highly dependent on international co-operation and confidence that if it turned out that the weather reports from various countries were being used for illegal actions the exchange of weather observations could be seriously prejudiced. Who wants to give weather observations to a country who is using them in warfare? For example, during World War II weather observations were secret and both the Germans and the Allies had their own networks of observation stations. The Germans constructed a system with automatic buoys along the Atlantic coast of Europe rising to the surface and then sinking down again every third hour.

All these possibilities of manipulating the weather which are gradually evolving demand wise consideration. In our age of more and more limited resources the struggle for the utilization of natural assets is of extreme importance. The struggle for the use of the oceans and fishing rights has been going on for many years and it now looks as if the use of the atmosphere may have to be regulated in the same way through international agreements.

9 MAN AND THE WEATHER

The weather affects the activities, the pleasures, the health, and indeed the safety of men and women almost every day of their lives – at home, during travel to and from school, shops, or place of work, and in the rest and recreation periods of evening or week-ends, especially if they involve such outdoor interests as sport or gardening.

It is for these reasons that most people have the daily habit of glancing at the weather forecasts in the newspapers or listening, perhaps with only half an ear, to the frequent forecasts provided on radio or television. More attention is paid when bad road conditions, due to heavy rain, ice, snow, fog or strong winds are expected. On such occasions the special road reports on the radio are most helpful, and many motorists will also ring up their local Weather Centre for advice, or dial for the recorded local area forecasts on the telephone.

Holidays are a special case, because bad weather can ruin the most carefully planned vacation; good holiday weather will help to provide that extra enjoyment which will last in the memory. It is not surprising that most of us seem to recall that the holidays of our youth were a long succession of perfect weather, for like a sun-dial, the memory only counts the sunny hours.

Our interest in the summer holiday weather starts well in advance, often in the middle of winter when plans are made to pick the best places for warmth and sunshine, the best times to explore the hills and mountains, or for small-boat sailing in sheltered waters, or more extensive cruising abroad.

In this chapter we shall take a closer look at the weather in different places and at different times of the year from several points of view, holidays, outdoor activities or travel, and will also touch upon the question of the effects of weather on our general health.

Weather for holiday makers

After a long cold dark winter many people wish to enjoy sunshine and warmth which will make swimming a pleasure, with little or no rain to spoil their outdoor enjoyment. To guarantee such 'perfect' weather, many choose to go to the Mediterranean, but it must be remembered that the height of summer, the school holiday period, is often too hot in such areas for those accustomed to a cooler climate. The earlier and later months (May or June, September or even October) still provide a degree of summer warmth in southern countries, and have the added advantage of avoiding the crowded conditions of high season.

The British holiday weather

The weather in the British Isles in summer can vary from the almost unbroken sunshine of 1976, when the absence of rain brought about a water supply crisis, to years when it seems to rain every other day. In fact rain, generally in the form of showers, is a common summer feature, which is why the countryside remains green and not a dried-up dusty yellow, and is why holiday resorts provide indoor entertainments such as concert or dance halls, a feature which is often absent in countries with drier summer climates. Temperatures are rarely uncomfortably excessive, and can be definitely on the cool side if the wind is blowing from a northerly direction.

A normal British summer is most suitable for the activity type of holiday, such as sailing, golf, riding, walking, climbing or fishing; anyone who wishes to be certain of enjoying sun-bathing 'bikini' weather would be well advised to seek a warmer drier climate further south or east. The seas around the British Isles are slow to warm up and swimming may be more of a challenge than a luxury. Hotels with heated swimming pools may not be exactly a necessity, but they are certainly a welcome amenity.

To most people, a summer holiday is one by the seaside, a habit which grew up in the last century and which shows little sign of being abandoned. It is thus fortunate that the areas in Britain most likely to have the most summer sunshine lie on the coasts. The places which regularly head the merit table of most hours of bright sunshine are those in the Isle of Wight and the Channel Islands, but all coastal resorts are sunnier than inland areas. In Ireland the south-east coasts are the most favoured, and in Scotland the east coast is sunnier than the west, although in May and early June, with a spell of easterly winds, the east coasts can be bitterly cold and overcast, while the west coasts and the Inner Hebrides can experience a sequence of cloudless days in atmospheres of almost unbelievable clarity.

Sunshine, however, does not always coincide with high temperatures. A day with north-westerly winds can be quite cool, and yet provide plenty of sun from a broken sky. On the other hand, a south-westerly day can be cloudy and yet bring a soft mild pleasant warmth. Everyone will agree that there is a difference between exhilarating or bracing weather and that which is relaxing or even oppressive, and yet it is extremely difficult to define the difference in a meteorological statistic, because the human reaction depends on a complex combination of temperature, sunshine, wind and humidity.

Because of the proximity to a relatively cool sea and the onset of afternoon sea breezes, the maximum summer temperatures tend to be higher inland than on the coast, but shelter from the wind can often counteract the effect of this numerical difference. In a hot summer, the highest recorded temperatures will always be found in those parts of Britain closest to the continent of Europe. In a normal or cool summer, the most clement conditions will be found further to the south-west, in Devon and Cornwall and along the southern coast of Ireland. The vegetation in these areas is evidence not only of the absence of hard winters, but also of the more dependable summer warmth.

The warmest summer month is almost invariably July or August, but pleasant holiday weather can also be experienced in May or June, and a good summer often extends into September. An 'Indian summer' in October can sometimes be enjoyed, but can hardly be depended upon.

The prevailing direction of the wind influences the nature and distribution of the best summer weather. With a dominant sequence of westerly winds, the eastern coasts are favoured, but when there is a dominance of easterly weather then the east coasts can be cloudy and cool while there will be fine sunny weather on the western side of the country. In 1968, when north-easterly winds brought cold wet weather to much of south-eastern England, southern Ireland had one of their best summers.

As far as rain is concerned, the driest part of Britain is beyond doubt the lowland and coastal areas of the east and south, but in a truly dry spell, rain is absent from all areas, even in the western hills where there is a high total of average annual rainfall. Southern Britain is more liable to dry spells than Scotland and the north, but when severe thunderstorms develop over southern England, often moving northward from France, then the northern areas can remain free from such extreme weather.

Thunderstorms, on average, are far more common in south-eastern England than elsewhere in Britain, giving rise to the cynical remark that an English summer is 'three fine days and a thunderstorm'. This is not to say that summer showers are less frequent in areas to the west and north, but merely that they are less severe.

Because of the variable nature of the British climate, which is such that its summer climatic statistics are more of a meteorological concept than a reliable guide, it is difficult to plan ahead with confidence for a holiday period at a particular time or particular place. Perhaps the best solution is to treat it as it comes, making short visits as the current weather dictates, remembering that in general the east is drier and the south is warmer; for a day trip it is always wise to travel downwind, like a locust swarm, in search of a pleasant destination, the only snag being that many other people may have the same sensible idea.

Weather in the hills and mountains

There is an appreciable change in climate in Britain from the coastal strip to inland districts, but an even more rapid change occurs when moving upward into the mountains. Temperature and sunshine decrease with an increase in height above sea level, while wind strength and rainfall amount and frequency increase. Weather changes during the day can be more rapid and those holiday-makers who seek the hills are more weather sensitive, and (if they are prudent) more weather conscious than those who prefer the seaside holiday.

Summer is of short duration on the higher hills and mountains, and even during this period of most favourable weather there can be hazards which cannot be ignored, for the mountains are no suitable playground for the inexperienced or the thoughtless.

The main potential dangers are fog and heavy rain. The fog is due to the cloud base being below the level of the high ground; cloud bases are always lower on the windward side of a hill, on the lee side the base is higher and the cloud more broken.

A hill fog can be a frightening experience to anyone unprepared for its sudden appearance, the total obliteration of all landmarks and consequent lack of any sense of direction. Good map reading and a reliable compass are essential to avoid becoming lost, stumbling into a bog, or possibly falling down a steep scree or precipice. Experienced hill walkers will make the most of such tactics as back-tracking, following the lead of stone walls, or the downward track of streams, but a beginner can be liable to panic and make stupid decisions. Before venturing into mountain regions, one should always listen to the local weather forecast, and be prepared to seek and take heed of experienced advice. Walking alone or with small children is not for the uninitiated and everyone, whether alone or in a group should always let someone know the plans for the day, so that search or rescue parties can be arranged if an emergency arises.

Heavy rains from summer storms can be very dangerous to anyone going pot-holing in the limestone hills. A safe dry underground passage can change to a raging torrent in a matter of hours or even minutes. Again, weather forecasts can help to assess the risks involved, and no one should go underground without letting others know.

Rock climbing is another sport in which the weather plays a significant part. Rain can quickly change the nature of a climb, and strong winds can greatly increase the dangers on an exposed cliff face. The lesson is clear; join a well-organized climbing club, and never ignore the weather prospects.

Adequate footwear and sensible clothing are essentials for activities on the heights. Weather conditions can change rapidly and there is a big difference between day and night weather. A combination of wet clothing, an increasing wind and sudden drops in temperature can create a major physical stress known as 'exposure'.

The combined effect of wind speed and temperature has been summarized in the following table:

Effective temperatures for moving air

Air temperature (°C)	+10	+5	0	−5	−10	−15	−20
Wind speed							
Calm	+10	+5	0	−5	−10	−15	−20
4·5 metres/sec.	+4	−2	−8	−14	−20	−26	−32
9 m/sec.	0	−7	−14	−21	−28	−36	−42
14 m/sec.	−2	−10	−17	−25	−33	−41	−48
18 m/sec.	−3	−11	−19	−27	−35	−43	−51
22 m/sec. (gale force)	−4	−12	−20	−28	−36	−44	−52

These effective temperatures are serious enough in themselves, but if the body is soaked to the skin, the resultant stress is even greater. When caught on the mountains at night it is important to keep as dry as possible and to seek the maximum shelter from the wind.

With the advent of autumn, weather conditions on the hills deteriorate rapidly and there is an increasing risk of snow. Visibility in cloud in a hill fog can be as much as twenty or thirty metres, but in snow it can be virtually nil, and the potential dangers are correspondingly far more serious.

Once snow has fallen in reasonable quantity on the mountains, the whole picture becomes transformed, and the area becomes the playground of the winter sports enthusiast.

Winter holidays

The use of skis for movement on foot in winter has been common practice for many ages in countries such as Scandinavia, but it was the English who introduced skiing as a sport in the Swiss Alps towards the end of the last century. 'The great popularity of wintersports,' writes Neville Lytton in his Preface to the volume Winter Sports in the Lonsdale Library, 'dates from the Grindelwald Conference. This was organised by Sir Henry Lunn in 1892; its object was to unite all the Christian churches; . . . Naturally, the Conference failed, but *à quelque chose malheur est bon* – Sir Henry noticed that mid-winter in the high Alps is an earthly paradise – the gloom and fog and damp are left in the valley below, while up in the eternal snows the sun is often too hot to bear and the sparkling atmosphere is more of a tonic than the finest champagne. He soon turned his organising ability to propagating this important discovery.' (From *Mountain Jubilee* by Arnold Lunn.)

During the last thirty years winter holidays have increased rapidly in popularity, and while the choice of winter resort is often determined by the personal choice of host nation, or by the availability of such amenities as ski-lifts or *après-ski* activity, some weather considerations are worth bearing in mind.

Snow will fall first in the early winter on the ski resorts at the highest levels, and will also remain there longer into spring than at sites lower down the mountains. During the winter snow reports are printed regularly in papers such as *The Times*. An enjoyable winter holiday needs both snow and sunshine, and sunny days are most likely at those sites well inland, although even there they cannot be guaranteed. Day length is important; in the north, in countries such as Norway or Sweden, the winter day is very short, being at times little over six hours, but further south some two to three more hours of daylight can be experienced. The northern resorts tend to come into their own late in winter or early spring, when the days are longer and yet the snow is still available. The greatest meteorological risk at a winter resort is that of avalanches.

Avalanches

The risk of avalanches increases in late winter. Avalanches are most easily formed along slopes with an inclination of between 20 and 40 degrees. Along the steep slopes snow usually slides down immediately after falling, but on gentle slopes there is not enough gravity to exert a pressure strong enough to make the snow start sliding and so snow masses mound up. Avalanches are caused when light snow falls on top of a layer of densely packed snow; a typical example is when frost snow falls on a layer of old packed snow. In spring the daily melting in the day, and freezing during the night of the top layer of snow creates a surface which is ideal for fresh snow to glide on.

Avalanches can be triggered by a lone skier. When the snow masses start moving they pile up at the foot of the slope, so that in narrow valleys the bottom can become completely filled with snow and be very dangerous for the skier. Skiers are advised to ask local people where the risky spots are and to observe the steep slopes carefully especially after periods with mild weather and fresh snow. In many places in the Alps avalanche warnings are issued regularly.

Holiday weather in southern Europe

In summer the sun blazes and the heat is intense in southern Europe and northern Africa. In July and August especially the heat is oppressive for 'northerners'. The map in Fig. 35 shows the holiday weather in central and southern Europe for the three summer months of June, July and August. The following information is given for each place shown on the map: the average afternoon temperature, the average lowest night-time temperature and the normal number of rainy days for each of the three summer months. The afternoon temperatures in central Europe are around +22°C. Temperatures a couple of degrees higher are found in France, Switzerland, southern Germany and Austria. The western coasts of Europe are often somewhat wetter than the inland areas, but the threat of summer rain is rarely serious south of Brittany.

South of the Alps, in the Mediterranean, we find the real summer heat. Afternoon temperatures are around 30°. On warm days they might be even higher; +40° is not uncommon. On Cyprus and in the inner parts of Spain +47° has been recorded, which is the heat record for Europe.

In the Mediterranean area the sea plays an important role in moderating the night-time temperatures. The lowest temperatures during the night are still numerically high, around 20°C which is the afternoon temperature of a northern summer, but they feel delightfully cool in the evening in contrast to the heat of the day. Later in the night, the fall in temperature brings about a rise in humidity, which can produce the sultry 'sticky' weather in which some people find it difficult to sleep.

Along the coasts of the Riviera, however, a steady sea breeze brings cooler air from the sea in over the beaches. The heat is deceptive and it is easy to get sunburnt if one doesn't take it easy in the sun for the first few days since the sun is very strong at these latitudes. The Mediterranean is characterized by winter rains and summer dryness and the number of rainy days during the summer is very low, less than two days of rain per month for most places. The sea also exerts a suppressing effect on convection and the formation of thunderstorms. Instead, they occur further inland especially in the mountains (in the Alps and Pyrenees). The maximum is over thirty-five days per year meaning that in summer there can be a thunderstorm nearly every second day. Most summer thunderstorms occur in the late afternoon or evening, but in central Europe it seems to thunder most frequently around 10 o'clock in the evening. In late summer when strolling along the Mediterranean beaches one may often see a string of flashing thunderstorms far out over the sea in the dark southern night.

The weather in August is very similar to that in July. In June, however, the dry, hot weather has not yet dried off or killed the vegetation so that the landscape has a fresher look and it is worth while considering spending a vacation in the Mediterranean in June rather than in the traditional months of July or August.

In winter the weather in the northern Mediterranean area can be quite gloomy and dull and even chilly. Snow may fall as far south as in Italy, Greece and Turkey. The Balearic Islands and the Spanish coasts stay fairly cool during the winter. However, many other places in North Africa, and in the Levant, have a degree of warmth even in winter. The table at the top of p. 252 shows the normal afternoon temperatures, night temperatures and

Fig. 35 Holiday weather in Europe and North Africa.
In the table opposite information is given about the normal afternoon temperature (mean maximum temperature), the normal lowest temperature during the night (mean minimum temperature) and the average, normal number of rainy days for the three summer months June, July and August. The places are shown on the map.

Place	Mean maximum temperature			Mean minimum temperature			Average normal number of rainy days		
	J	J	A	J	J	A	J	J	A
Stockholm	19	22	90	10	14	13	8	9	10
London	20	23	22	10	13	12	11	13	13
Amsterdam	20	21	21	10	12	12	15	14	19
Berlin	21	23	22	10	13	12	9	10	10
Warsaw	23	24	23	12	15	14	7	11	7
Paris	23	24	24	11	13	13	8	11	10
Venice	25	27	27	17	19	18	8	7	7
Rome	28	31	31	15	18	18	3	2	3
Geneva	23	25	24	13	14	14	10	8	9
Nice	24	27	27	16	19	19	2	1	1
Lisbon	24	27	28	15	17	18	2	1	1
Madrid	26	30	30	14	16	16	4	2	1
Athens	29	32	32	19	22	22	2	1	1
Istanbul	25	27	28	16	18	19	6	3	4
Vienna	21	24	23	13	15	14	9	9	10
Malaga	26	29	29	19	21	22	0	0	1
Palma de Mallorca	26	29	30	16	19	20	2	0	1
Split	26	30	30	17	20	19	6	3	3
Agadir (Morocco)	24	26	27	16	18	18	1	0	1
Rhodes	26	28	28	21	23	24	1	0	0
Tunis	29	32	32	18	20	21	3	1	2
Algiers	27	28	29	18	21	22	2	0	0
Funchal	22	24	25	17	19	20	1	0	0
Las Palmas	24	26	26	18	19	21	1	0	0
Tel Aviv	27	29	30	18	20	21	0	0	0
Rimini	26	28	28	17	19	19	7	4	5
Alghero	25	28	28	16	18	18	2	1	1
Varna	26	30	29	16	19	18	9	6	6
Constanta	24	27	27	16	18	19	8	5	4
Marrakesh	31	36	36	17	20	20	2	0	1
Palermo	27	30	30	18	21	21	2	0	2

the average number of rainy days during the rest of the year for some places around the Mediterranean and in the Canary Islands.

It is evident from the table that temperature varies very little in the Canary Islands. In January the afternoon temperature goes down to +21°C and the night-time temperature to 14°C which means a good chance of sunshine and warmth; the number of

Place	September			November			January			April		
Rhodes	27	22	1	20	16	7	15	10	13	19	15	4
Tunis	29	20	5	20	12	10	15	7	13	21	11	7
Las Palmas, Canary Islands	26	20	1	24	18	7	21	14	6	22	16	3
Tel Aviv	29	19	1	24	13	7	18	8	10	21	12	2
Palma de Mallorca	27	18	5	18	10	7	14	5	7	19	9	4
Marrakesh (Morocco)	32	18	1	21	10	6	18	7	7	24	12	6
Palermo	28	19	4	21	12	8	16	8	12	20	41	6

rainy days increases somewhat to six per month. Messina and Palma de Mallorca are fairly cool in winter and rain falls almost every third day. Also Rhodes is cool with rain almost every second day. In April, however, spring is back, the weather is less unsettled and is considerably warmer and drier, comparable with the summer weather of northern Europe.

Sailing weather

Those who 'go down to the sea in ships' have always been conscious of the vagaries and dangers of the weather, and wind and tide have to be given serious consideration. Rear-Admiral Sir Francis Beaufort (1774–1857), of the British Navy, was probably the first to interpret the performance of a sailing ship in terms of wind strength.

His original wind specification, laid down in 1805, was as follows:

Beaufort Number	Wind Description	Specification	
0	Calm.	Calm.	
1	Light air.	Just sufficient to give steerage way.	
2	Slight breeze.	That in which a well-conditioned man-of-	1–2 knots
3	Gentle breeze.	war with all sail set and 'clean full' would go in smooth water	3–4 knots
4	Moderate breeze.	from:	5–6 knots
5	Fresh breeze.	That which she could	Royals, etc.
6	Strong breeze.	just carry in chase 'full and by':	Single-reefed top-sails or top-gallant sails.

Beaufort Number	Wind Description	Specification
7	Moderate gale.	Double-reefed topsails, jib, etc.
8	Fresh gale.	Triple-reefed topsails, etc.
9	Strong gale.	Close-reefed topsails and courses.
10	Whole gale.	That which she could scarcely bear with close-reefed main topsail and reefed foresail.
11	Storm.	That which would reduce her to storm staysails.
12	Hurricane.	That which no canvas could withstand.

(Note that Force 7 is no longer called a moderate gale, but only a high or strong wind; the term gale is restricted to Forces 8 and above.)

For use at land stations observing the weather, the specifications were later modified as follows:

Beaufort Number	Specification	Wind strength miles per hour	Wind strength metres per second
0	Calm; smoke rises vertically.	less than 1	less than 0·3
1	Direction of wind shown by smoke drift, but not by wind vanes.	1–3	0·3–1·5
2	Wind felt on face; leaves rustle; ordinary vane moved by wind.	4–7	1·6–3·3
3	Leaves and small twigs in constant motion; wind extends light flag.	8–12	3·4–5·4
4	Raises dust and loose paper; small branches are moved.	13–18	5·5–7·9
5	Small trees in leaf begin to sway; wavelets form on inland waters.	19–24	8·0–10·7
6	Large branches in motion; whistling heard in telegraph wires; umbrellas used with difficulty.	25–31	10·8–13·8
7	Whole trees in motion; inconvenience felt when walking against the wind.	32–38	13·9–17·1
8	Breaks twigs of trees; generally impedes progress.	39–46	17·2–20·7
9	Slight structural damage occurs (chimney pots and slates removed).	47–54	20·8–24·4
10	Seldom experienced inland; trees uprooted; considerable structural damage occurs.	55–63	24·5–28·4

Beaufort Number	Specification	Wind strength miles per hour	metres per second
11	Very rarely experienced; accompanied by widespread damage.	64–72	28·5–32·6
12		Above 73	Above 32·7

(Note that the wind ranges quoted refer to the speed over a period of time. Gusts of short duration can produce speeds of up to twice this mean value.)

It is to the everlasting credit of Admiral Beaufort that this classification is very similar to the one in use today. Shipping forecasts, and reports from coastal stations broadcast for the guidance of ships at sea outside the sheltered coastal waters, still use the Beaufort scale for the indication of the strength of the wind.

In the area around the British Isles, such forecasts are issued as follows:

On 200 kHz (1500 m) at 00.33, 06.33, 13.55 (11.55 on Sundays), and 17.55 hours. The areas covered are Southeast Iceland, Faroes, Bailey, Rockall, Shannon, Fastnet, Sole, Finisterre and Biscay in the Atlantic; Viking, Forties, Fisher, Dogger and German Bight in the North Sea; Fair Isle, Cromarty, Forth, Tyne, Humber, Thames, Dover, Wight, Portland, Plymouth, Lundy, Irish Sea, Malin and Hebrides around the British coasts. Gale warnings are also issued on this wavelength, as and when issued; they are also added, when appropriate, to the regular forecasts. These warnings are issued when mean winds of at least Force 8 or with gusts reaching 43 knots are expected. The term 'severe gale' implies mean winds of at least Force 9 or gusts reaching 52 knots. The term 'storm' implies a mean wind of at least Force 10, or gusts reaching at least 61 knots. The term 'imminent' implies within 6 hours of the time of issue; 'soon' implies between 6 and 12 hours; 'later' implies more than 12 hours.

These shipping forecasts are planned by international agreement, so that each country is responsible for a particular area of the sea or ocean. They are designed for the larger vessels, from the super-tankers to the type of privately-owned yacht or motor-boat which is cruising on the Irish Sea, the North Sea or the Bay of Biscay.

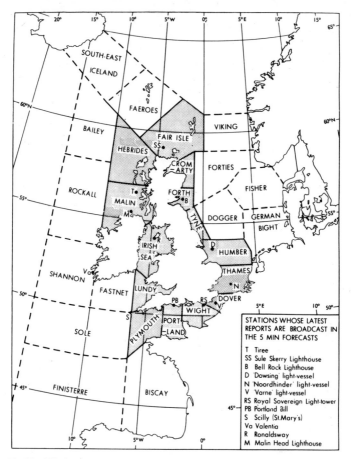

Fig. 36 Shipping forecast areas around the British Isles (reproduced by permission of the Controller of Her Majesty's Stationery Office).

Special forecasts are prepared in Britain for smaller boats which sail in the close inshore relatively sheltered waters. Details are as follows:

Forecasts for 'Inshore Waters' are designed to give as much detail as possible over the sea up to 12 miles (about 20 km) offshore. They are broadcast by the B.B.C. early every morning and again late at night, usually just before close-down. The late broadcast also gives 22.00-hour reports from selected coastal stations. There are also some twenty-two Post Office and Irish Coastal Radio stations which broadcast weather forecasts and gale warnings on VHF wave-bands at 08.03 and 20.03 hours.

Small boat sailing

Small boat sailing has increased greatly in popularity in recent years, and is still regarded by many as a seasonal sport, being restricted in winter by the coldness of the water and by the threat of strong winds and gales.

The seas, owing to their large specific heat, may remain warmer than the land during winter, but they are very slow to warm up with the coming of spring. Winter sailing on inshore or inland waters has been made more feasible by the introduction of warmth retaining wet-suits, but it is still principally a summer pastime; the winter being a time for refitting and repainting boats.

The coldness of the sea, especially in spring and early summer, is a cause of one of the major hazards in coastwise sailing, namely sea fog. If warm moist air moves over a cold sea surface, it is cooled, and cloud forms on the surface. This sea fog has the additional hazard of distorting sound waves, so that fog sirens or hooters can appear to originate from completely the wrong directions. Most cruising yachts now carry some form of radar to overcome this difficulty, but sea fog is still a danger to small craft, especially if they are near the shipping lanes.

Wind is a necessity for sailing, but it can still be a great danger if it becomes too plentiful. Luckily the summer season is one of generally lighter winds than the winter half-year. Few small boats are safe to handle in winds which exceed Force 5, except in very

Fig. 37 The influence of islands, narrows and promontories on the wind.
The wind is greatly affected by terrain. These illustrations show what happens to the wind when it blows past different geographical features. The motion of the air is shown by means of stream lines, being closest together where the wind is strongest. The wind blows in the direction of the arrows. The influence is greatest when the air is stably stratified, which is often the case at sea or along the coast. Narrows channel the flow of air and give rise to particularly strong winds. The wind also becomes stronger around points or promontories.

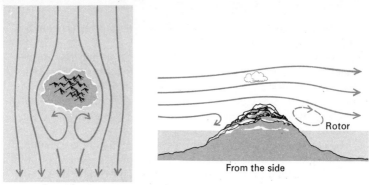

From above

From the side

Air flow around an island at stable stratification

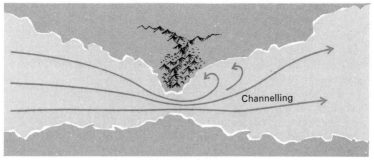

Channelling

Air flow through a strait

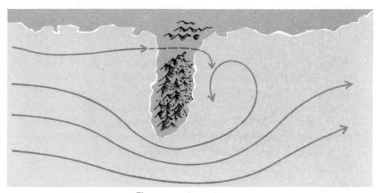

Flow around a point

experienced hands, and this is why Force 6 is often referred to as the 'Yachtsman's Gale'.

There is no such thing as a perfectly steady wind, the movement of the air near the surface being constantly changeable about the mean values both in speed and direction. This variability is most obvious on the sea coast and is affected by the nature and shape of the coast, the presence of nearby islands or cliffs and the stability or instability of the air mass.

In hot summer weather with a weak general wind circulation, the situation can be dominated by the (day-time) sea breeze, from sea to land, and the night-time land breeze in the reverse direction. The sea breeze extends up to 30–40 km inland from the coast, and is confined to the lowest 2–400 m of atmosphere. A typical sea breeze is about 5 m/sec., but the land breeze is lighter. The sea breeze often starts in late morning or early afternoon blowing at right angles to the coastline, but it veers clockwise during the day, under the influence of the earth's rotation. In colour plate 'Winds I–II', these local winds are discussed in further detail.

The gustiness and short-time variations in the wind will be greatest when the air is unstable, or in other words when showers or large cumulus clouds are present. In a heavy storm such gusts and rapid changes of wind direction in the immediate vicinity of the storm can be dangerous to small boats, especially during the passage of cold fronts or minor 'line squalls'.

The skilled helmsman can, however, take advantage of the gentler type of gust if he remembers that as the wind dies away it backs in direction; as it strengthens in the gust it veers anything up to a couple of compass points, dying away and backing again after the gust is over.

Thus, when beating on the starboard tack, the helmsman will bear away slightly as the wind drops, but luff smartly to point higher at the peak of the gust, then bearing away again. By this process he can often steal many yards to windward over his less observant rivals. On the port tack, the technique is more difficult, because the luffing must take place in the calmer period before the gust arrives; at the height of the gust the boat has to be laid off, but as soon as it is over, the boat can again be pointed higher. On a reach, the same advantage can be gained by careful management of the jib or mainsheets.

Winds do not only affect the sails, they influence the waves. When it has been blowing steadily from one direction for some time, waves are produced on the sea surface. The size and period of the waves depend on the wind speed, its duration, the distance to the shore line and the depth of the sea. When the waves come in over shallow water, they tend to grow and the horizontal wavelength becomes larger. In so doing they become unstable and commence to break, which is why offshore shoals can often be detected by breaking waves.

It takes time for the wave to build up to its maximum height. With a wind of about 5 m/sec., the time needed is only about 5 hours, but with a 15 m/sec. wind, some two days are required to reach the maximum corresponding wave height. The following table is an approximate indication of the height of fully developed waves corresponding to a given wind speed in an inland sea such as the Baltic.

Wind speed	Wave height
2–5 m/sec.	0·5 m
7·5–10	1·2
12·5–15	2·1
22–25	3·5

Uncomfortable states of the sea can occur when the wind is blowing against the tide, a state of affairs often encountered in the North Sea. In the Atlantic, a different type of problem arises when a heavy swell is generated by strong winds blowing far away at the other side of the ocean.

A small boat sailor faces many weather problems, but with care and experience most of them can be overcome, provided that not too many risks are taken.

Motoring weather

The weather can create a lot of troubles for car drivers. Many accidents are directly due to bad weather conditions ranging from slippery roads to bad visibility at night. Even if bad weather conditions are not always responsible for car accidents the motorist still has to consider the weather when planning a trip. In snow and fog and when the roads are slippery one has to drive slowly and cautiously and the car should have been prepared appropriately for the weather. Car trips take longer and traffic jams easily form.

Each season has its own type of difficulty, but usually summer is the time when we expect the least adverse weather on the road. Certainly the influence of weather on car driving is at a minimum during summer as compared to the winter half of the year.

Rain storms, thunderstorms and hail

Nevertheless, in summer the cumulonimbus cloud, often accompanied by thunderstorms, heavy rain and hail, can cause considerable trouble for the motorist. Even if the car is the safest place in a thunderstorm, a thundercloud can produce large amounts of rain in a very short time. Rain amounts exceeding 25 mm in half an hour is by no means uncommon. In heavy rain the wipers may not be efficient enough to keep the windscreen clear of water. Visibility becomes bad and the best thing to do is to park at the side of the road and wait for the heaviest rain to pass. Many people are irritated by lightning and find it difficult to concentrate on driving. The heavy rain from a cumulonimbus cloud normally lasts only for a short time, of the order of half an hour, so there is little risk of a long delay in waiting for the rain to end.

Thunderstorms are often connected with hail. The size of hailstones varies, but in most cases in Europe they are too small to cause damage. In exceptional cases hailstones can grow big enough to cause considerable damage to a moving car. Hail is made of ice particles which can cause temporary slippery road conditions so that it is again wise to stop and wait for the hail storm to pass.

During and after heavy rain the road surface can be very wet. At high speeds a phenomena known as 'hydroplaning' can occur; when braking at high speed the wheels may get locked and the car lifted up on the thin film of water on the road so that it glides as on water skis on a water surface. The effect is similar to ice on the road – the driver loses control over the vehicle and an accident may result. So, if there is a lot of water on the road – slow down!

This has the added advantage of causing less spray of water behind the car which can blind other drivers.

We normally get the worst kind of weather for driving in winter. We all know there is going to be another winter every year with slippery roads, snow storms and bad visibility, but it looks as if many drivers are surprised every year when the first frost makes the roads icy and the first heavy snowfall starts. The main weather-induced factors that affect us in winter can be treated under three headings:

Reduced visibility
Slippery road conditions
Other traffic hazards

Reduced visibility

Reduced visibility can arise in several different ways. The commonest one is, however, in connection with fog. In dense fog, visibility falls below 50 metres.

How different types of fog form is described in the colour plate 'Fog' and on page 79.

Radiation fog is most frequent in autumn when it forms during the evening and night because of radiational cooling of the ground. The fog favours low levels of ground and thickens in hollows and in marshy terrain. It can occur in patches meaning that when driving we go in and out of fog all the time, which can be very dangerous.

In late autumn and early winter advection fog gets more frequent. It covers large areas and the variations in visibility from place to place are fairly small. Advection fog can occur both at night and in the day, especially during a temporary thaw of snow-covered land.

Visibility can be reduced in other ways than by fog formation. Snowfall, mist and drizzle are other types of weather associated with low visibility. In winter it is, in addition, dark most of the time which certainly adds to the strain. Rain, sand, salt and sleet make the windscreen and headlights dirty, reducing visibility even further. In such cases it is better to stop and clean the lights and the windscreen. It is also better to prepare oneself for bad road conditions by always checking that the car is in good shape before starting a trip, and that the liquid container for the screen washers is filled.

When driving in fog these actions do not help much, however. The small droplets in the fog scatter the light beams from the headlights and a major fraction is reflected back towards the driver. If driving with full headlights the light is directed higher up and the scattered and reflected light is quite strong. The fog appears as a thick white 'porridge' in front of the driver. It is then better to use a dipped beam instead, which is directed more along the surface of the road. Less light gets reflected and it becomes easier to see the road. However, the only safe way is to cut down one's speed.

Slippery road conditions

The first frost is usually the start to a new season of difficult conditions for the car driver. In a couple of hours in the morning the first glazed frost causes hundreds of cars to go off the road and breakdown vans have some busy hours ahead. In Britain the first autumn frost usually occurs about the middle of October, but is rarely severe so early in the season.

It is not always easy for the weather service to forecast the first frost giving slippery roads. The following three points should give some guidance to the observant motorist.

1. When humid air settles during a clear and calm night it is time to be careful. If fog and mist have formed during the night and the temperature is around freezing point the risk for frosty roads is high. It is also worth remembering that the temperature at the surface can be below freezing even if a thermometer a couple of metres above the ground shows temperatures above freezing.

2. Another common case causing slippery roads is when the sky is clearing after a day of rain giving wet road surfaces. A rapid clearing in the evening may cause a rapid temperature drop and freezing on the road before the water has run off or evaporated.

3. Other dangerous situations are slippery road conditions caused by freezing rain or drizzle. Freezing fog does not only

reduce the visibility, but depositing frost on the surface gives slippery road conditions.

In winter most accidents with slippery roads or other traffic obstructions occur in connection with snow and temperatures around freezing. Freezing rain is most likely to cause extremely bad road conditions – it can be more like driving on a skating rink than on a road. Situations of that kind occur when very warm air streams towards the north after a long period with very cold weather. The freezing rain consists of raindrops which have a temperature below freezing. On meeting the ground they immediately freeze giving a continuous ice sheet on the road. Rain at temperatures above freezing can also change to ice on the road if it previously has been cooled by a long period of cold temperatures. In practically all cases of slippery roads the conditions can be more or less patchy. Particularly dangerous places are bridges, low-lying road sections, where the cold air accumulates more easily, and stretches facing north to which the sun hardly reaches in winter. In spring and late winter, snow and ice melt during the day and melting water runs out over the road. In the night the water freezes again giving rise to patches of ice along the road.

Other traffic hazards

Heavy snowfalls, particularly in connection with strong winds, do not only cause slippery roads, but may make the roads impossible to negotiate. The strong wind piles up the snow into drifts which are impossible to penetrate. Naturally in a case like this there is nothing to do but stop or leave the car at home, or even better, stay indoors.

Strong winds can also create hazardous conditions along exposed bridges or roads, across open country or along the coast. In extreme cases the car can be blown off the road.

What can we do then in order to safeguard ourselves against all the risks to which the weather exposes us? The first thing to do, of course, is to slow down and drive carefully. In winter in snowy areas we should switch to winter tyres, and in any case make sure that the tyre treads are in good condition.

It is also advisable to look at the rest of the equipment of the car; check that the wipers work properly and that the container is filled with anti-freeze liquid. In winter it is a good idea to carry a snow shovel and a box with sand in the car.

Of course we should listen to the daily weather forecasts. With the frequent issue of radio and television forecasts we should never be completely caught out on the road or surprised by the weather. The local radio stations issue traffic advice when bad weather is expected. They also usually furnish more detailed information about road conditions in the district.

Slippery roads may develop though even if the weather service has not issued a warning. It is really up to oneself to check what state the road is in. Take it easy!

Weather and health

The weather has a strong effect on our comfort. A rapid change in weather influences working efficiency and many diseases get worse when cold and unsettled weather sets in.

Many people claim that they can anticipate an approaching storm by feeling pains in joints, scars or by increased arthritic pains. A headache before a thunderstorm is not uncommon. How the body reacts to these changes in advance of the arrival of a storm is not known and very few investigations have been performed to find this out. It is not impossible to imagine that changes in the field of atmospheric electricity, normally creating a difference in potential of 100 volts over a metre, can affect our nerve cells. The few investigations carried out, however, seem to indicate that we actually are very insensitive to all kinds of electromagnetic radiation.

On the other hand we are not as ignorant when other more obvious aspects of weather are concerned. The human body works at a temperature close to $37°C$. At a temperature of $28-30°C$ and calm winds the body is in balance with the environment. No perspiring or shivering is needed to keep the body at a constant temperature. The surrounding air temperature, however, as we well know, can change quite a lot both indoors and outdoors. This is also true for humidity. In order to adjust to the annual variation of temperature we dress differently in summer and winter. We also heat our houses in winter and if living in a hot climate we use air conditioning in summer. In other cases, despite these precautions, we have difficulty in adapting to the surrounding climate and the body reacts by perspiring to transport the heat away from the skin or by shivering to increase the production of heat by mechanical work.

Humidity

If the humidity rises our chances of keeping down the body temperature by perspiring diminish, so that the air feels muggy and sticky. If there is a wind ventilation increases, more air passes over the skin, and heat is transported away more effectively, so that it is more comfortable and we can more easily endure warm and moist air. The summer heat in northern Europe is usually dry and pleasant while the climate in southern latitudes is considerably more humid. At times every summer the north is invaded by air masses from the south, causing great discomfort with hot and moist air.

In winter at temperatures below zero the air is dry in an absolute sense, even if the relative humidity is 100%. In that case it is difficult to maintain a comfortable humidity level indoors. Normal and pleasant humidity at 20°C is about 60%. If we have a ventilation and heating system which takes in air from the outside heating it to +20° the relative humidity can get very low, well below 35%, at which level we start to feel definite discomfort with an irritated throat and coughing. In many heating systems moisture is therefore added to the intake air before distribution to the rooms.

Heat waves and air pollution

Air pollution has become an integral part of weather in the big urban areas. High amounts of air pollutants have a harmful effect on many diseases. In the U.S.A. and Japan, where air pollution often reaches high levels during calm hot weather, many groups of sensitive people are badly affected. Particularly exposed are those people with heart troubles, allergies and respiratory diseases.

Diabetes patients are particularly affected during prolonged hot spells.

The first real heat wave of the year in the U.S.A. usually causes more deaths than later ones in summer even if the temperatures then are higher – being unused to strong and prolonged heat seems to play a major role. During a heat wave it is usually not the hot days that are the most strenuous, but the high night-time temperatures denying the body a recovery during the night. The level of air pollution normally increases very rapidly during hot weather in America's big cities such as Los Angeles, New York and Washington D.C. The wind is light and the air becomes practically impossible to breathe. Warnings are issued to old people to stay indoors and sensitive people are advised to have the air conditioning switched on.

The extreme conditions experienced in the U.S.A. are very rarely experienced in Europe, but the strain on the body increases even in more moderate heat spells, especially in southerly winds.

Weather and working efficiency

Our sensitivity to changes in the weather does not only show up in variations in illnesses. Working efficiency, reactions and the number of accidents are all affected markedly when changes in weather occur. In Germany an extensive investigation has shown that the number of traffic accidents depend more on people's reaction to weather changes than the traditional weather factors such as slippery roads, low visibility, snowfall and strong winds. Types of weather which have a bad influence on people are the passage of a low pressure system, warm and humid weather after a warm front and the cold winds in the rear of a depression. Such types of weather are called 'biologically unfavourable'. In Hamburg the number of traffic accidents increased up to 40% in such biologically unfavourable weather conditions while the increase of accidents in connection with glazed frost was only 6·4% and in connection with fog 5·2%. On average in the whole investigation the number of traffic accidents increased by 10% in biologically unfavourable weather conditions.

There are not only biologically unfavourable types of weather, but also favourable ones. The clearing of clouds behind a low or stable high pressure weather gives a marked increase in working efficiency. Another German study shows that working efficiency may go up by 20–35% in biologically favourable weather situations. Generally we become more energetic in such periods of weather.

Weather and diseases

Experience shows that many diseases can get worse or improve as weather changes. In some cases, however, we may exaggerate the effects of weather; the so-called 'spring depression' that some people say they are affected by depends less on the spring than on other factors that happen to coincide with the arrival of spring. Naturally we feel that rainy and cloudy weather is dull and gloomy, but that does not necessarily mean that it affects our mental health. The suicide rate does not vary either by season or by changes in weather. On the other hand there are clear connections between the variation of weather and a number of other diseases. The following table is a summary of some common diseases and how they are influenced by variations in weather:

Short-periodical effects	*Long-periodical, or seasonal, effects*
Asthma	
Increases with sudden cooling, especially connected with strong gusty winds. Foggy or humid weather usually makes the asthma patient feel better.	Low in winter, suddenly increasing after June. Maximum in late autumn.
Bronchitis	
Increasing complaints during fog and in polluted air. A change to cold weather makes conditions worse.	High in winter, low in summer.

Short-periodical effects	Long-periodical, or seasonal, effects
Skin cancer More common with increasing number of sunshine hours. Increased risks for white people living in the tropics.	
Glaucoma (acute) Most attacks during very cold days in winter or very warm days in summer.	Maximum in November, less frequent in summer.
Conjunctivitis (acute) Common in sunny weather.	Common in May and September.
Rheumatic diseases Most forms of arthritis react to strong cooling or falling temperature in connection with strong winds. Humidity does not seem to have any effect. Some patients react to strong heat.	Complaints particularly common in autumn and early winter.
Heart diseases: Infarctions and Angina Pectoris Occur more frequently shortly after a period of strong cooling.	Highest mortality occurs in January–February. In warm, southern countries conditions seem to be reversed.
Common cold Despite all that has been said on this subject an American investigation shows that it is commonest in cold weather, which is also true for coughs.	Maximum in winter.

Short-periodical effects	*Long-periodical, or seasonal, effects*
Peptic Ulcer	
Ulcers are aggravated when drastic changes in temperature occur, for example, in changes of air masses with a weak predominance for a change to colder weather.	Least frequent in mid-summer, maximum in winter.
Epilepsy	
Increased number of attacks in connection with cold fronts being followed by deep cold air.	Maximum during November and December. Minimum in summer.

The right column is a summary of statistical investigations of the frequency of the yearly variation of different diseases. It does not mean that a causal relation is established between the annual variation of weather and the frequency of a particular disease. The seasonal variations of diseases may be due to causes other than the weather.

INDEX